SUSTAINABLE CONSTRUCTION PRACTICES FOR INDUSTRY 4.0

Steven Smith

Wisdom Publisher

ISBN: 9798852905208
Imprint: Independently published

Cover design by: Art Painter
Library of Congress Control Number: 2018675309
Printed in the United States of America

To all the visionary builders and creators, whose commitment to sustainability and innovation shapes a brighter and greener future for our world. May your endeavors inspire generations to come.

CONTENTS

INTRODUCTION

In the ever-evolving landscape of construction and design, a new era of innovation and responsibility has emerged - sustainable construction. This groundbreaking approach to building is transforming the industry, presenting a compelling vision of eco-consciousness and forward-thinking. As we confront pressing global challenges like climate change, resource depletion, and urbanization, the imperative for sustainable construction has never been greater.

This book sets out on a journey through the realm of sustainable construction, delving deep into the array of practices that pave the way for a more environmentally friendly and socially responsible built environment.

We will traverse the vast landscape of sustainable construction materials, uncovering the wonders of bamboo - nature's engineering marvel, the insulating and fire-resistant qualities of cork, the strength and durability of recycled steel, and the potential of reclaimed wood to add character and sustainability to projects. This journey will also unveil the promise of innovative materials like hempcrete, ferrock, and recycled plastic - each contributing unique advantages to sustainable construction.

In the quest to create eco-friendly structures, sustainable construction is not limited to material choices alone. We will also delve into the crucial principles and considerations of sustainable design, energy-efficient building techniques, water conservation strategies, waste reduction, and indoor environmental quality.

Armed with this knowledge, professionals in the construction industry can create spaces that not only minimize their environmental impact but also enhance occupants' health and well-being.

The power of technology in sustainable construction cannot be ignored, and the author will explore how digital tools, Building Information Modeling (BIM), and Internet of Things (IoT) applications are reshaping the way we build, optimizing processes, and maximizing resource efficiency.

The journey will also take a turn towards the circular economy - the pinnacle of sustainability, where the emphasis is on reusing, recycling, and upcycling materials to create a regenerative construction ecosystem. From sourcing materials responsibly to transporting them to the construction site and optimizing their installation, we will uncover sustainable procurement practices and construction techniques that further strengthen the circular economy's foundation.

CHAPTER 1: INTRODUCTION TO SUSTAINABLE CONSTRUCTION

1.1 Understanding Sustainable Construction

S ustainable construction is a critical approach to building that aims to minimize negative environmental impacts while maximizing resource efficiency. It involves adopting practices that consider the entire lifecycle of a structure, from planning and design to construction, operation, maintenance, and eventually, deconstruction. This approach goes beyond merely constructing environmentally friendly buildings; it involves a comprehensive and holistic view of the entire construction process.

One of the key aspects of sustainable construction is environmental responsibility. This entails reducing the carbon footprint of buildings by employing energy-efficient technologies and practices. It also involves choosing materials that have a lower environmental impact and implementing strategies to conserve energy and water resources. Sustainable construction seeks to minimize the negative impact that buildings have on the natural environment.

Resource efficiency is another fundamental pillar of sustainable construction. The goal is to make the most efficient use of resources, such as water and raw materials, to minimize waste generation and resource depletion. This includes adopting practices that promote recycling, reusing materials, and reducing unnecessary consumption. By optimizing resource usage, sustainable construction contributes to a more sustainable future.

Central to sustainable construction is the concept of green building design. This approach prioritizes energy efficiency, natural ventilation, and the use of renewable energy sources. Green buildings are designed to provide a healthier and more comfortable living or working environment while minimizing energy consumption and greenhouse gas emissions. Integrating green design principles into construction projects can lead to significant environmental benefits.

The choice of materials used in construction is crucial to sustainability. Sustainable construction emphasizes the use of environmentally friendly materials that have a lower impact on the environment. This includes materials that are responsibly sourced, have a reduced carbon footprint, and are recyclable or biodegradable. By opting for sustainable materials, the construction industry can play a pivotal role in reducing its overall environmental impact.

Sustainable construction begins in the project's design phase, but even general contractors can make a big difference by choosing better materials. By adopting sustainable building processes today, construction companies can make a significant difference for both the environment and their clients. The construction industry's impact on the environment is substantial, with estimates suggesting that it's responsible for more than 30% of global natural-resource extraction and nearly half of the world's carbon emissions.

Despite the challenges, the construction industry is making progress towards sustainability. Executives in the engineering and construction sectors have increasingly prioritized sustainability in the design phase, recognizing the importance of sustainable practices in the construction process. Government regulations, cost-efficiency considerations, and expanding environmental concerns are influencing sustainable change in the industry.

To achieve sustainability goals, construction companies are working towards creating a more sustainable supply chain while addressing cost-efficiency issues. Sustainable design, circular economy principles, and renewable energy adoption are some of the strategies being implemented to reduce the industry's environmental impact. While there's still much work to be done, the shift towards sustainable construction is an important step towards a greener and more sustainable future for the built environment.

1.2 Importance of Eco-Friendly Building Techniques

Eco-friendly building techniques play a crucial role in addressing the pressing environmental challenges faced by the construction industry and the planet as a whole. These sustainable practices are essential for mitigating the negative impact of construction activities on natural resources, energy consumption, and carbon emissions.

Eco-friendly building techniques prioritize resource efficiency and environmental responsibility throughout a building's lifecycle. By incorporating sustainable materials, renewable energy sources, and green construction processes, these techniques significantly reduce the environmental footprint of buildings. This reduction in waste generation, resource extraction, and energy consumption helps conserve natural resources and protect ecosystems.

Traditional construction methods are a significant contributor to greenhouse gas emissions, which exacerbate climate change. By adopting eco-friendly building techniques, construction companies can lower carbon emissions and contribute to global efforts to combat climate change. Sustainable construction practices focus on energy-efficient designs, alternative energy sources, and carbon-neutral building materials, making them instrumental in achieving a more sustainable and resilient built environment.

Eco-friendly building techniques also promote healthier indoor environments for occupants. Sustainable materials with low volatile organic compounds (VOCs) improve indoor air quality, reducing the risk of respiratory issues and other health problems. Additionally, natural ventilation, efficient insulation, and green building designs contribute to better thermal comfort and overall well-being for building occupants.

Adopting eco-friendly building techniques can provide economic benefits in the long run. While sustainable materials and construction methods may require a higher initial investment, they lead to reduced operating costs over the building's lifetime. Energy-efficient buildings consume less electricity and water, resulting in lower utility bills for owners and tenants. Sustainable buildings often command higher property values and rental premiums, making them financially attractive investments.

Eco-friendly building techniques are of paramount importance for the construction industry and society at large. By reducing the environmental impact, lowering carbon emissions, promoting healthier indoor environments, and providing economic benefits, these practices pave the way for a more sustainable and resilient future. Embracing eco-friendly building techniques is not only a responsible choice but also a necessary step towards building a better world for future generations.

1.3 Environmental Impacts of the Construction Industry

The construction industry has significant environmental impacts that extend across various stages of a building's lifecycle. From resource extraction to waste generation, these impacts can have far-reaching consequences on the environment and contribute to pressing global challenges such as climate change and resource depletion.

One of the primary environmental impacts of the construction industry is its heavy reliance on natural resources. Construction activities require vast amounts of raw materials, including sand, gravel, timber, and minerals. The extraction of these resources can lead to habitat destruction, soil erosion, and disruption of ecosystems, affecting biodiversity and ecological balance.

Energy consumption is another major environmental concern associated with construction. The building process, including manufacturing materials and operating heavy machinery, is energy-intensive and heavily reliant on fossil fuels. This reliance contributes to greenhouse gas emissions, which are a significant driver of climate change.

The construction industry is a major generator of solid waste. Construction sites produce immense amounts of debris, including unused materials, packaging, and demolition waste. Improper disposal of construction waste can lead to pollution of land and water bodies, impacting both human and environmental health.

Construction activities also contribute to air pollution. Dust and particulate matter released during construction can degrade air quality and pose health risks to workers and nearby communities.

Additionally, construction vehicles emit pollutants, such as nitrogen oxides and volatile organic compounds, further deteriorating air quality.

Land use change is another critical environmental impact of the construction industry. As urbanization expands, natural landscapes are transformed into built environments, leading to habitat loss and fragmentation. This encroachment on natural habitats disrupts wildlife and biodiversity, threatening the survival of various plant and animal species.

Water consumption is a significant concern in construction, particularly in regions facing water scarcity. Construction processes require substantial amounts of water for mixing cement, cooling equipment, and other purposes. The excessive use of water resources can exacerbate water stress in already water-limited areas.

The transportation of construction materials and equipment contributes to greenhouse gas emissions and air pollution. The long-distance transportation of materials from suppliers to construction sites adds to the carbon footprint of construction projects.

Addressing the environmental impacts of the construction industry is essential for sustainable development and environmental conservation. Adopting eco-friendly building techniques, using sustainable materials, promoting energy efficiency, and reducing waste generation are crucial steps towards minimizing the industry's negative environmental footprint. Sustainable construction practices can lead to more resilient buildings and infrastructure that contribute to a greener and healthier future for both the planet and its inhabitants.

1.4 The Role of Materials Management in Sustainable Construction

Materials management plays a crucial role in promoting sustainable construction practices and mitigating the environmental impact of the construction industry. Effective materials management encompasses various strategies aimed at optimizing resource use, reducing waste generation, and promoting the adoption of eco-friendly materials throughout a building's lifecycle.

One of the key aspects of materials management in sustainable construction is the careful selection of materials. Sustainable construction prioritizes the use of environmentally friendly and renewable materials, such as recycled content, reclaimed wood, and low-impact composites. By choosing these materials, construction projects can reduce their carbon footprint and minimize the depletion of finite resources.

Materials management also involves the responsible sourcing of materials. Sustainable construction practices emphasize obtaining materials from ethical and eco-conscious suppliers. Ensuring that materials are sourced sustainably helps prevent deforestation, habitat destruction, and other environmentally harmful practices associated with resource extraction.

Effective materials management also involves reducing waste generation at construction sites. By implementing efficient inventory control and precise ordering, construction companies can minimize excess material that often ends up as construction waste. Proper on-site handling and storage of materials further prevent damage and spoilage, maximizing their usage.

Recycling and reusing materials are integral to sustainable materials management. By recycling construction waste and repurposing materials from demolition or renovation projects, the industry can divert substantial amounts of waste from landfills and reduce the demand for new raw materials.

Pre-fabrication and modular construction techniques are also significant components of sustainable materials management. Off-site construction reduces material waste and energy consumption, while modular components allow for more efficient assembly and disassembly, promoting circular economy principles.

Beyond the construction phase, materials management extends to the maintenance and eventual deconstruction of buildings. Designing for deconstruction ensures that materials can be easily disassembled and reused or recycled at the end of a building's life, contributing to a more sustainable and circular approach to construction.

Incorporating Building Information Modeling (BIM) technology in materials management enhances efficiency and sustainability. BIM enables real-time tracking of materials, waste, and energy consumption, facilitating better decision-making and reducing the environmental impact of construction projects.

Effective materials management in sustainable construction is a collaborative effort involving architects, contractors, suppliers, and other stakeholders. By embracing eco-friendly materials, responsible sourcing practices, waste reduction strategies, and innovative technologies, the construction industry can transition towards a more sustainable future, reducing its ecological footprint and promoting environmental conservation.

CHAPTER 2: SUSTAINABLE MATERIALS FOR CONSTRUCTION

2.1 Overview of Sustainable Construction Materials

Sustainable construction materials play a vital role in reducing the environmental impact of the construction industry and promoting a greener built environment. These materials are carefully selected based on their eco-friendliness, resource efficiency, and potential for reuse or recycling. By using sustainable construction materials, the industry can contribute to conservation efforts, reduce greenhouse gas emissions, and minimize waste generation.

One of the key categories of sustainable construction materials is recycled and reclaimed materials. These materials are obtained from post-consumer and post-industrial waste streams and repurposed for construction purposes. Examples include recycled concrete, reclaimed wood, and recycled steel. By using these materials, construction projects can divert waste from landfills and reduce the need for virgin resources.

Another important category of sustainable construction

materials is rapidly renewable resources. These materials come from sources that can replenish quickly, such as bamboo, cork, and certain types of timber. Utilizing rapidly renewable materials helps preserve natural ecosystems and reduces the environmental impact associated with resource extraction.

Sustainable construction also emphasizes the use of low-impact materials with a smaller ecological footprint. Low-impact materials include products with reduced emissions during manufacturing and those that require fewer resources to produce. For instance, low-VOC (volatile organic compound) paints and adhesives contribute to improved indoor air quality.

The concept of embodied carbon is crucial in sustainable construction materials. Embodied carbon refers to the carbon emissions associated with the entire lifecycle of a material, from extraction to disposal. Sustainable construction materials aim to minimize embodied carbon by using low-carbon alternatives and by adopting circular economy principles that encourage recycling and reuse.

Biodegradable materials are gaining traction in sustainable construction. These materials can decompose naturally at the end of their useful life, reducing waste and environmental pollution. For example, biodegradable insulation materials made from natural fibers offer an eco-friendly alternative to traditional synthetic insulation.

High-performance materials with enhanced durability are essential for sustainable construction. Durable materials require less frequent replacement, resulting in reduced resource consumption and waste generation over time. Utilizing high-performance materials also contributes to the overall longevity and resilience of constructed assets.

Innovative technologies and material science advancements continue to expand the range of sustainable construction materials available. These include developments in green concrete, energy-efficient insulation, and sustainable composites. By embracing these innovations, the construction industry can make significant strides toward achieving more sustainable building practices.

It is essential to consider the full life cycle of sustainable construction materials, from sourcing and manufacturing to use and eventual disposal. A life cycle assessment allows for a comprehensive evaluation of a material's environmental impact, enabling better-informed decisions in materials selection.

Overall, sustainable construction materials form the foundation of eco-friendly building practices. By prioritizing recycled and reclaimed materials, rapidly renewable resources, low-impact alternatives, and biodegradable options, the construction industry can contribute to a more sustainable and environmentally responsible built environment. Embracing innovative technologies and conducting life cycle assessments further enhances the industry's ability to achieve a greener future.

2.2 Bamboo: A Versatile and Renewable Building Material

Bamboo is a versatile and renewable building material that has gained significant attention in sustainable construction practices. It offers numerous advantages, making it an attractive alternative to traditional building materials. Bamboo is a fast-growing grass that can reach maturity in just a few years, making it highly renewable and sustainable compared to slow-growing timber species.

One of the key benefits of bamboo is its exceptional strength-to-weight ratio. It has a higher tensile strength than many types of wood and even some steel. This makes bamboo a suitable choice for structural applications in construction, such as beams, columns, and flooring. Its strength and flexibility make it resistant to wind, earthquakes, and other environmental stresses. Bamboo is also highly adaptable and can be used in a variety of construction techniques. It can be used as solid bamboo poles or engineered into composite materials such as bamboo plywood, laminated bamboo lumber, or bamboo fiberboards. These engineered bamboo products offer enhanced dimensional stability, improved strength, and durability.

Another advantage of bamboo is its rapid growth and ability to sequester carbon dioxide. As bamboo grows, it absorbs a significant amount of carbon dioxide from the atmosphere, making it an effective tool for carbon sequestration. Additionally, bamboo forests can help mitigate soil erosion and contribute to biodiversity conservation.

Bamboo is a low-maintenance material that requires minimal chemical treatments compared to traditional timber. It has natural resistance to pests, fungi, and fire, reducing the need for chemical treatments and improving its overall sustainability profile. Additionally, bamboo can be harvested without killing the entire plant, allowing for regrowth and sustainable harvesting practices.

The aesthetic appeal of bamboo adds to its popularity in sustainable construction. Its natural grain patterns and warm colors create a visually pleasing environment. Bamboo can be used in various architectural elements, including flooring, wall cladding, furniture, and decorative finishes.

In addition to its environmental benefits and versatility, bamboo offers economic advantages. It is a cost-effective material, particularly in regions where it is locally available. The abundance and fast growth of bamboo make it an affordable option for construction projects, particularly in developing countries.

However, there are considerations when using bamboo in construction. Proper harvesting techniques, treatment methods, and sourcing from responsibly managed bamboo forests are crucial to ensure its sustainability. It is essential to work with reputable suppliers and follow established standards and certifications for sustainable bamboo products.

Sourcing Bamboo: Sustainable Bamboo Farming Practices

Bamboo has emerged as a popular and eco-friendly building material in sustainable construction due to its remarkable properties and rapid growth rate. However, to fully embrace the benefits of bamboo and minimize its environmental impact, it is essential to adopt sustainable bamboo farming practices. Sustainable bamboo sourcing involves responsible cultivation and harvesting techniques, ensuring the longevity of bamboo forests and preserving natural ecosystems.

One of the key aspects of sustainable bamboo farming is the selective harvesting of mature bamboo culms. Unlike trees, which are often cut down entirely for timber, bamboo can be selectively harvested, allowing the remaining culms to continue growing and promoting natural regeneration. This practice ensures that the bamboo forest remains intact, maintaining its ability to absorb carbon dioxide and support biodiversity. Sustainable harvesting also prevents soil erosion and protects watersheds, making bamboo farming an environmentally friendly alternative to traditional logging.

To cultivate bamboo sustainably, farmers must use organic and natural farming practices. Bamboo is naturally pest-resistant, reducing the need for pesticides and chemical fertilizers. Organic farming methods promote soil health, biodiversity, and water conservation, making bamboo cultivation a low-impact agricultural activity. By adopting these practices, bamboo farmers can ensure that their bamboo forests thrive while minimizing negative environmental effects.

Sustainable bamboo farming often involves adhering to

internationally recognized certification standards. Organizations like the Forest Stewardship Council (FSC) and the Sustainable Forestry Initiative (SFI) provide certification to bamboo products sourced from responsibly managed forests. These certifications offer consumers and construction companies assurance that the bamboo used in their projects comes from sustainable sources and adheres to strict environmental and social criteria.

Sustainable bamboo farming practices not only benefit the environment but also support local communities. Bamboo cultivation can provide a source of income for rural communities, creating employment opportunities and economic development. Additionally, bamboo forests can serve as natural barriers against soil erosion and flooding, protecting nearby communities from natural disasters. The economic and ecological value of bamboo cultivation contributes to the overall sustainability of the construction industry.

Utilization in Construction: Structural Elements, Flooring, Wall Panels

Bamboo's versatility and exceptional mechanical properties make it suitable for a wide range of construction applications. As a construction material, bamboo can be utilized in structural elements, flooring, and wall panels, showcasing its adaptability and strength in various building components.

In structural applications, bamboo demonstrates remarkable strength-to-weight ratio properties, comparable to steel. Bamboo beams, columns, and trusses have been successfully used to support roofs, floors, and walls in buildings. Its high tensile strength and flexibility make bamboo an excellent choice for earthquake-prone regions, where it can withstand seismic forces and contribute to resilient building designs. Bamboo's rapid regrowth rate ensures a steady supply of building materials for continuous construction projects, reducing the demand for timber from natural forests.

Bamboo flooring has gained popularity as a sustainable and aesthetically pleasing alternative to traditional hardwood flooring. The process of creating bamboo flooring involves slicing bamboo into strips, which are then laminated together to form sturdy and visually appealing floorboards. The finished product exhibits an elegant and unique grain pattern, providing a natural and warm ambiance to interior spaces. Moreover, bamboo flooring's eco-friendly characteristics align with green building principles, contributing to healthier indoor air quality and reduced environmental impact.

Apart from flooring, bamboo is widely used in the construction of wall panels and cladding. Bamboo wall panels offer both functional and aesthetic benefits. They serve as excellent insulators, enhancing the building's energy efficiency and reducing heating and cooling costs. Additionally, bamboo's natural beauty and distinctive appearance add an elegant touch to both interior and exterior spaces. The use of bamboo wall panels in construction contributes to sustainable building practices and the promotion of eco-friendly designs.

Incorporating bamboo into construction projects aligns with the principles of circular economy and sustainability. By opting for bamboo as a building material, construction companies can reduce their environmental footprint and support eco-friendly practices. Bamboo's rapid growth rate and minimal environmental impact make it an ideal choice for a more sustainable built environment.

Sourcing bamboo responsibly through sustainable farming practices and utilizing it in construction projects for structural elements, flooring, and wall panels exemplify the importance of eco-friendly building techniques. The adoption of sustainable bamboo sourcing ensures the preservation of bamboo forests and their invaluable contribution to carbon sequestration and biodiversity. Bamboo's versatility and strength make it a suitable substitute for traditional building materials, promoting resilient and sustainable building designs. Embracing bamboo in construction not only reduces the construction industry's environmental impacts but also supports the well-being of local communities and fosters a more sustainable and greener future for the built environment.

2.3 Hempcrete: Lightweight and Eco-Friendly Construction Solution

Hempcrete, also known as hemp-lime or hemp concrete, is a lightweight and eco-friendly construction solution gaining recognition in sustainable construction practices. It combines the natural fibers of hemp with a lime-based binder to create a bio-composite material that offers numerous advantages for both the environment and building occupants. As a renewable and low-carbon material, hempcrete presents a compelling alternative to conventional construction materials, promoting energy-efficient buildings with minimal environmental impact.

The production of hempcrete begins with the cultivation of industrial hemp plants, which are known for their fast growth and minimal need for pesticides or herbicides. Hemp crops absorb significant amounts of carbon dioxide from the atmosphere as they grow, making them effective carbon sinks. This process, known as carbon sequestration, contributes to reducing greenhouse gas emissions, making hempcrete a carbon-negative material. Additionally, hemp plants have deep root systems that enhance soil stability and prevent erosion, further supporting sustainable agricultural practices.

After harvesting, the hemp stalks are separated from the leaves and seeds, leaving behind the fibrous inner core called the hurd. The hurd is then mixed with a lime-based binder, typically composed of lime and water. This combination forms a lightweight and insulating material with excellent thermal and acoustic properties. Hempcrete is not load-bearing but is used as an infill material within timber or steel frames, providing both structural stability and thermal insulation.

One of the most significant benefits of hempcrete lies in its insulating properties. Hempcrete's porous structure allows it to regulate moisture and temperature, creating a comfortable and healthy indoor environment. It effectively controls humidity levels, reducing the risk of mold and mildew growth. The insulation provided by hempcrete reduces the need for mechanical heating and cooling, leading to energy savings and lower carbon emissions over the building's lifespan.

Hempcrete is also a breathable material, allowing water vapor to pass through the walls without trapping moisture. This feature contributes to improved indoor air quality, preventing the buildup of harmful pollutants and allergens. As a non-toxic and non-allergenic material, hempcrete is suitable for those with chemical sensitivities or respiratory issues, promoting a healthier living space.

Hempcrete's fire-resistant properties add an extra layer of safety to buildings, making it suitable for both residential and commercial applications. The lime binder in hempcrete acts as a natural fire retardant, reducing the spread of flames and limiting fire damage. This feature contributes to enhanced building safety, particularly in areas prone to wildfires or where fire-resistant materials are required by building codes.

The use of hempcrete aligns with circular economy principles as it represents a closed-loop system. At the end of a building's life, hempcrete can be recycled or repurposed as an agricultural soil amendment, returning valuable nutrients to the soil and closing the loop of sustainability. This recyclability further reduces waste and promotes a more sustainable building industry.

Hempcrete stands as a lightweight and eco-friendly construction solution that exemplifies the importance of sustainable building materials. Its renewable nature and carbon-negative characteristics make it a compelling alternative to traditional construction materials, contributing to reduced greenhouse gas emissions and supporting sustainable agricultural practices. The insulation and breathable properties of hempcrete create comfortable and healthy indoor environments while reducing energy consumption and promoting indoor air quality. Additionally, its fire-resistant properties enhance building safety and resilience. By incorporating hempcrete into construction projects, the construction industry can take significant strides towards achieving a more sustainable and environmentally conscious future.

Sourcing Hempcrete: Obtaining Hemp Shives and Lime-Based Binder

The sourcing of materials is a crucial aspect of sustainable construction, and the same principle applies to hempcrete. Hempcrete is composed of two primary components: hemp shives and a lime-based binder. Let's delve into the process of sourcing these materials sustainably.

Hemp shives, or hemp hurd, are the inner core of the hemp stalks and serve as the primary aggregate in hempcrete. The cultivation of industrial hemp plants is necessary to obtain high-quality hemp shives. Industrial hemp is known for its rapid growth and minimal need for pesticides or herbicides, making it an environmentally friendly crop. Hemp crops also have the unique ability to sequester carbon dioxide from the atmosphere as they grow, thereby reducing greenhouse gas emissions. Additionally, hemp plants have deep root systems that help prevent soil erosion and improve soil health. These characteristics make hemp cultivation a sustainable and regenerative practice.

Once the hemp plants have matured, they are harvested, and the stalks are processed to separate the hemp shives from the outer fibers and leaves. This separation can be achieved using mechanical decortication methods, which involve breaking down the stalks and mechanically removing the hurd. Mechanical decortication is a low-energy process that minimizes waste generation and can be performed without the use of harsh chemicals. This ensures that the sourcing of hemp shives is environmentally responsible and aligned with sustainable practices.

The lime-based binder used in hempcrete is typically composed of hydrated lime, which is produced through the heating of limestone, and water. Limestone is a naturally abundant mineral that can be extracted sustainably from quarries. It is important to source lime from quarries that adhere to responsible mining practices, minimizing the ecological footprint associated with extraction. Lime production itself has a relatively low environmental impact compared to other building materials. The lime binder acts as the adhesive that binds the hemp shives together, creating the hempcrete matrix.

Utilization in Construction: Insulation, Wall Systems, Sustainable Masonry

Hempcrete offers a wide range of applications in sustainable construction, including insulation, wall systems, and sustainable masonry.

One of the primary uses of hempcrete is as insulation material. The porous nature of hempcrete allows it to provide excellent thermal insulation properties. The trapped air within the hempcrete matrix creates a barrier against heat transfer, reducing energy consumption for heating and cooling. The insulation properties of hempcrete help maintain a comfortable indoor environment by minimizing temperature fluctuations and reducing the reliance on mechanical heating and cooling systems.

This results in energy savings and a reduced carbon footprint for the building.

Hempcrete is also commonly used in wall systems. It can be applied as an infill material within timber or steel frames, providing both structural stability and insulation. When used in wall construction, hempcrete acts as a breathable material, allowing the walls to regulate moisture levels. This breathability helps prevent the buildup of moisture within the walls, reducing the risk of mold and improving indoor air quality. Additionally, hempcrete's acoustic properties contribute to sound insulation, creating quieter indoor spaces.

In sustainable masonry, hempcrete can be used as an alternative to traditional masonry materials like bricks or concrete blocks. The lightweight nature of hempcrete makes it easier to handle and reduces the structural load on the building. It can be cast into various shapes and sizes, allowing for flexibility in design. Hempcrete masonry offers excellent thermal insulation properties and contributes to energy-efficient buildings. Moreover, hempcrete's fire-resistant properties enhance the safety of the structure.

Beyond its applications in insulation, wall systems, and masonry, hempcrete can also be used for other construction elements such as floors, roofs, and partitions. Its versatility and eco-friendly characteristics make it an attractive choice for sustainable construction projects.

The sourcing of hempcrete materials, such as hemp shives and lime-based binders, should align with sustainable practices, including environmentally responsible cultivation and responsible mining of limestone. Hempcrete finds extensive utilization in sustainable construction, particularly in applications such as insulation, wall systems, and sustainable masonry. Its thermal insulation properties, breathability, and fire resistance contribute to energy-efficient and comfortable buildings. Additionally, hempcrete's lightweight nature and versatility make it suitable for various construction elements. By incorporating hempcrete into construction practices, the industry can promote sustainable building solutions that minimize environmental impact and prioritize energy efficiency.

2.4 Cork: Insulation and Fire-Resistant Material

Cork is a remarkable material that serves as an excellent choice for insulation and fire-resistant applications in the construction industry. It possesses a range of unique properties that make it highly suitable for eco-friendly building practices.

Cork is sourced from the bark of the cork oak tree (Quercus suber), primarily found in Mediterranean regions like Portugal, Spain, and North Africa. The cork harvesting process is conducted in an environmentally conscious manner, where skilled workers carefully remove the outer bark without harming the tree. This sustainable method allows the tree to regenerate its bark, ensuring its continued health and longevity. Subsequent cork harvests take place every 9 to 12 years, providing a continuous and renewable supply of cork.

In terms of insulation, cork offers exceptional thermal properties. Its cellular structure, consisting of tiny air-filled pockets, acts as an effective barrier against heat transfer. This natural insulation ability helps to keep buildings cool in hot climates and retain warmth in colder regions. By incorporating cork insulation into construction projects, builders can improve energy efficiency and reduce heating and cooling demands.

Cork's insulating properties are not limited to temperature control. It also possesses excellent acoustic insulation properties, making it an ideal material for noise reduction within buildings. The cellular structure of cork absorbs sound vibrations, preventing them from traveling through walls and floors. This makes cork an effective solution for creating quiet and peaceful environments, whether it's in residential, commercial, or public spaces. Additionally, cork's acoustic insulation benefits contribute to enhanced privacy, better concentration, and improved overall comfort.

Another significant advantage of cork is its fire resistance. Due to its natural composition, cork is highly resistant to combustion and does not emit toxic gases when exposed to fire. Instead, it smolders slowly without bursting into flames, acting as a fire retardant. This inherent fire-resistant property makes cork a safe material choice for construction applications, particularly in areas where fire safety is a priority. Incorporating cork insulation can help enhance the overall fire performance of a building, providing valuable time for occupants to evacuate and reducing the spread of flames.

Cork is also an eco-friendly and sustainable material. The cork oak forests, from which cork is sourced, play a crucial role in biodiversity conservation and carbon sequestration. These forests are home to various plant and animal species and contribute to mitigating climate change by absorbing carbon dioxide from the atmosphere. By utilizing cork in construction, builders can support the preservation of these valuable ecosystems and reduce their environmental impact.

Sourcing Cork: Harvesting from Cork Oak Trees

Cork, a natural and sustainable material with numerous applications in the construction industry, is primarily sourced from the bark of cork oak trees (Quercus suber). These trees are predominantly found in Mediterranean regions, such as Portugal, Spain, and North Africa. Cork harvesting is a delicate and environmentally conscious process that ensures the continued health and regeneration of cork oak trees. Skilled workers carefully strip the outer bark from the trees without causing harm, allowing them to regrow new bark for future harvests. This sustainable harvesting method has been practiced for centuries, making cork a renewable and eco-friendly material choice.

The process of harvesting cork begins when the trees reach a certain age, typically around 25 years old. The first harvest, known as the "virgin cork," is not suitable for commercial use due to its irregular and rough appearance. Subsequent harvests, which occur every 9 to 12 years, yield high-quality cork used in various applications, including construction.

Utilization in Construction: Flooring

One of the most popular applications of cork in construction is as flooring material. Cork flooring is known for its unique blend of durability, comfort, and eco-friendliness. It is manufactured using cork granules or tiles that are carefully bonded together with resins or adhesives. The resulting flooring material is soft underfoot, providing a comfortable surface for walking and standing for extended periods.

Cork flooring's inherent resilience allows it to bounce back from indentations caused by furniture or foot traffic, making it a long-lasting and low-maintenance flooring solution. Additionally, cork flooring possesses natural acoustic insulation properties, reducing noise transmission between floors and creating a quieter indoor environment.

Another advantage of cork flooring is its thermal insulation capability. The cellular structure of cork traps air, making it an efficient barrier against heat transfer. As a result, cork flooring helps maintain a comfortable indoor temperature and can contribute to energy savings by reducing the need for additional heating or cooling.

Also, cork flooring is hypoallergenic and resistant to mold and mildew, making it an excellent choice for individuals with allergies or respiratory sensitivities. Its natural properties also make it resistant to insects and pests, further enhancing its suitability for various construction projects.

Utilization in Construction:
Wall Coverings

Cork's versatility extends to wall coverings, where it offers both aesthetic and functional benefits. Cork wall coverings are available in various forms, such as cork tiles or cork panels, which can be applied directly to walls for a natural and visually appealing finish.

Cork wall coverings add a touch of warmth and texture to interior spaces, enhancing the overall ambiance of a room. The natural variations in cork's appearance create a unique and organic look, complementing different design styles and color schemes.

Apart from its aesthetic appeal, cork wall coverings also serve practical purposes. Similar to cork flooring, cork on walls provides acoustic insulation, reducing sound transmission and creating a quieter living or working environment. The material's ability to absorb sound vibrations makes it ideal for spaces where noise reduction is a priority, such as offices, libraries, or residential areas.

Cork wall coverings also contribute to thermal insulation, helping maintain a comfortable indoor temperature and reducing energy consumption. Additionally, cork's natural fire-resistant properties add a layer of safety to interior spaces, making it an ideal choice for commercial buildings and public facilities.

Utilization in Construction:
Sound Insulation

Cork's acoustic properties make it a favored material for sound insulation applications in construction. Cork can be used as a standalone sound barrier or combined with other materials to enhance acoustic performance in buildings.

As a standalone material, cork is often used to create acoustic panels or partitions that absorb sound and reduce noise transmission between rooms. These panels can be installed in various locations, such as office cubicles, conference rooms, or home theaters, to create quieter and more private spaces.

In addition to its sound-absorbing capabilities, cork is also used as an underlayment for flooring materials like hardwood or laminate. By placing cork underlayment between the subfloor and the flooring material, sound vibrations and impact noise are minimized, resulting in a quieter and more comfortable living or working environment.

Cork's lightweight nature and easy installation further contribute to its popularity in sound insulation applications. Its eco-friendly properties and sustainable sourcing make it an attractive choice for builders and architects seeking environmentally responsible construction materials.

2.5 Straw Bales: Energy-Efficient and Renewable Resource

Straw bales, traditionally considered agricultural waste, have gained recognition as a sustainable and energy-efficient resource in the construction industry. Composed of the dry stalks of crops such as wheat, rice, barley, or oats, straw bales offer numerous benefits as a building material, including excellent insulation properties, renewable sourcing, and minimal environmental impact.

Utilization in Construction: Insulation

One of the primary advantages of straw bales in construction is their exceptional insulation capabilities. The hollow structure of straw bales traps air, creating a natural barrier against heat transfer. This results in high thermal resistance, keeping buildings warm in winter and cool in summer. As a result, straw bale buildings often require less energy for heating and cooling, contributing to reduced greenhouse gas emissions and energy consumption.

When stacked and tightly compacted, straw bales form thick walls that provide superior insulation compared to conventional materials. These walls exhibit low thermal conductivity, effectively preventing heat from escaping or entering the building. The natural insulating properties of straw bales offer an energy-efficient solution for sustainable construction, aligning with green building principles and energy conservation goals.

Utilization in Construction: Wall Systems

Straw bales can serve as load-bearing components in wall systems, where they act as the primary structural elements of a building. These load-bearing straw bale walls eliminate the need for additional framing, reducing the consumption of other materials like wood or steel. As a result, straw bale construction can significantly reduce the overall carbon footprint of a building.

To construct load-bearing walls with straw bales, the bales are stacked in a running bond pattern and held in place with wooden stakes or reinforced with natural fibers, such as bamboo or hemp, for added stability. Once the walls are in place, they can be plastered with a clay or lime-based render to provide protection from the elements and create a smooth, finished appearance.

Straw bale wall systems offer excellent structural integrity and durability when built and maintained properly. Additionally, the thickness of the walls and their inherent insulating properties contribute to a comfortable and consistent indoor climate, making them ideal for both residential and commercial construction projects.

Utilization in Construction: Sustainable Masonry

Straw bales can be combined with other construction materials to create sustainable masonry that meets modern building standards. For instance, straw bale walls can be used as infill material within a timber or steel frame, providing insulation while reducing the overall use of conventional materials.

To create straw bale infill walls, the straw bales are placed between the structural elements of the frame, and then a plaster or stucco finish is applied to the exterior and interior surfaces. This method allows for the incorporation of straw bales into conventional building practices without compromising on structural integrity or aesthetics.

Straw bale infill walls contribute to the energy efficiency of a building by enhancing its insulation, thereby reducing the need for mechanical heating and cooling. Additionally, using straw bales as infill material promotes sustainable practices by utilizing a readily available agricultural byproduct that would otherwise be discarded.

Sourcing Straw Bales: Agricultural Residue

The sourcing of straw bales is an essential aspect of their sustainability. As an agricultural byproduct, straw is readily available after the harvest of cereal crops like wheat, rice, barley, or oats. By using straw bales in construction, builders can contribute to the reduction of agricultural waste and its environmental impact.

Straw bales are sourced from fields where cereal crops have been harvested. During the harvesting process, the stalks of the plants are collected and bundled into bales for easy transport and storage. These straw bales are then used in construction projects, reducing the need for other insulation materials that may have a higher environmental footprint.

Agricultural residue like straw is a renewable resource, as crops are replanted and harvested each year. By using straw bales in construction, builders can support sustainable agricultural practices and help foster a circular economy by transforming waste into valuable building material.

2.6 Recycled Plastic: Repurposing Waste for Construction

Recycled plastic has emerged as a promising and innovative material in the construction industry, offering a sustainable solution to repurpose waste and reduce environmental impacts. By transforming discarded plastic into durable building products, recycled plastic contributes to waste reduction, energy conservation, and a more circular economy.

Utilization in Construction: Building Components

Recycled plastic is utilized in various building components, providing a range of sustainable alternatives to traditional materials. One common application is in the manufacturing of plastic lumber, also known as recycled plastic timber. Plastic lumber can replace conventional wood in applications such as decking, fencing, and outdoor furniture.

Unlike traditional wood, plastic lumber does not require cutting down trees, preserving forests and conserving natural resources. It is also resistant to rot, insects, and water damage, resulting in longer-lasting and low-maintenance building components. Besides, by incorporating recycled plastic lumber into construction projects, builders can prevent plastic waste from ending up in landfills or polluting oceans.

Utilization in Construction: Insulation

Recycled plastic can also be transformed into insulation materials for buildings, providing an eco-friendly alternative to traditional insulation products. One example is thermal insulation made from recycled plastic bottles. The bottles are processed into polyester fibers, which are then used as insulation material to trap heat and reduce energy loss in buildings.

Recycled plastic insulation offers excellent thermal performance and moisture resistance, helping maintain a comfortable indoor environment while reducing energy consumption for heating and cooling. Additionally, using recycled plastic for insulation diverts plastic waste from landfills and contributes to a more sustainable approach to building design.

Utilization in Construction: Green Roofs and Green Walls

Green roofs and green walls are sustainable design elements that incorporate vegetation into building structures. Recycled plastic plays a role in these features by being used as a base material for green roof trays or wall panels.

Green roof trays made from recycled plastic provide a lightweight and modular solution for creating green roofs on top of buildings. These trays can be filled with soil and planted with vegetation, promoting biodiversity, reducing stormwater runoff, and improving insulation. By using recycled plastic for green roof trays, construction projects contribute to waste reduction and support urban greening initiatives.

Similarly, recycled plastic panels are used to create green walls, where vegetation can grow vertically on the building facade. These green walls improve air quality, provide thermal insulation, and add aesthetic value to the building. Incorporating recycled plastic into green wall systems demonstrates a commitment to sustainability and innovative building practices.

Sourcing Recycled Plastic: Plastic Waste Collection and Sorting

The sourcing of recycled plastic begins with the collection and sorting of plastic waste. Recycling facilities, municipal recycling programs, and waste management companies play a crucial role in gathering plastic waste from various sources, such as households, businesses, and industrial facilities.

Once collected, the plastic waste is sorted based on its type and quality. Different types of plastic are separated to ensure that only suitable materials are used for recycling. Advanced sorting technologies, including automated systems and optical scanners, help streamline the process and improve the efficiency of plastic waste sorting.

Sourcing Recycled Plastic: Recycling and Processing

After sorting, the plastic waste undergoes recycling and processing to transform it into usable materials. This process typically involves cleaning, shredding, and melting the plastic waste to create pellets or granules. These recycled plastic pellets can then be used to manufacture a wide range of building products, including plastic lumber, insulation materials, and green roof trays.

Recycled plastic products must meet quality and safety standards to ensure their suitability for construction applications. Rigorous testing and quality control measures are implemented during the manufacturing process to guarantee that the recycled plastic materials perform well and meet industry standards.

In conclusion, recycled plastic provides a sustainable solution for repurposing waste in the construction industry. From building components to insulation, and even in green roofs and walls, recycled plastic demonstrates its versatility and environmental benefits. The sourcing of recycled plastic involves efficient waste collection, sorting, and recycling processes that contribute to waste reduction and a more circular economy. By incorporating recycled plastic into construction projects, builders can contribute to a greener and more sustainable future.

2.7 Recycled Steel: Strength and Durability with Minimal Environmental Impact

Recycled steel has become a fundamental element in sustainable construction, offering exceptional strength, durability, and minimal environmental impact. By reusing scrap metal and transforming it into high-quality construction materials, recycled steel plays a crucial role in reducing the demand for new steel production, conserving natural resources, and curbing carbon emissions.

Sourcing Recycled Steel: Scrap Metal Recycling and Processing

The sourcing of recycled steel begins with scrap metal recycling and processing. Steel is one of the most recycled materials globally, and its recycling process is highly efficient. The primary sources of scrap steel include end-of-life vehicles, discarded appliances, demolished buildings, and industrial waste.

Recycling facilities and scrap yards collect and sort scrap steel, ensuring that contaminants are removed and the steel is segregated based on its grade and composition. Advanced technologies, such as shredders and magnetic separators, aid in breaking down and sorting the scrap metal efficiently.

Once the scrap steel is sorted, it undergoes a melting process in electric arc furnaces or basic oxygen furnaces. These furnaces can reach extremely high temperatures, melting the scrap steel and transforming it into liquid steel. The liquid steel is then cast into various shapes, such as bars or sheets, to create construction-grade steel products.

Utilization in Construction: Reinforcement

One of the primary applications of recycled steel in construction is as reinforcement in concrete structures. Reinforced concrete combines the compressive strength of concrete with the tensile strength of steel, creating a robust and durable building material.

Recycled steel rebar, also known as reinforcing bars, is widely used to reinforce concrete in various construction projects, such as buildings, bridges, and highways. The rebar provides additional strength to the concrete, enhancing its ability to withstand tension and preventing cracks and structural failures.

Using recycled steel rebar in construction significantly reduces the environmental impact associated with new steel production. It conserves energy, raw materials, and water, while also lowering greenhouse gas emissions and diverting steel waste from landfills.

Utilization in Construction: Structural Frameworks

Recycled steel is also utilized in the construction of structural frameworks for buildings and other structures. Steel's high strength-to-weight ratio makes it an ideal choice for supporting heavy loads and creating large open spaces in buildings.

Steel beams, columns, and trusses made from recycled steel are commonly used to create the skeleton of a building. These structural elements provide stability and support to the overall structure, ensuring it can withstand various loads and environmental forces.

By using recycled steel in structural frameworks, construction projects contribute to sustainability by reducing the need for new steel production and conserving natural resources. Additionally, steel's durability and long lifespan make it a cost-effective and environmentally responsible choice for construction materials.

Environmental Benefits of Recycled Steel in Construction

The utilization of recycled steel in construction offers several significant environmental benefits. Firstly, recycling scrap steel reduces the demand for raw materials, such as iron ore and coal, which are required for traditional steel production. This conserves natural resources and reduces the environmental impact of mining and extraction activities.

Secondly, the recycling process for steel consumes less energy compared to the production of new steel. This results in lower greenhouse gas emissions and contributes to the reduction of carbon footprints in construction projects.

Also, recycling steel waste prevents it from ending up in landfills, reducing the volume of waste and minimizing the risk of soil and water pollution. It also curtails the need for waste disposal and landfill management, saving valuable space and resources.

2.8 Reclaimed Wood: Sustainable Timber for Building Projects

Reclaimed wood has emerged as a popular and sustainable choice in the construction industry, offering a unique combination of environmental benefits and aesthetic appeal. By salvaging and reusing old wood from various sources, construction projects can reduce the demand for new timber, conserve natural resources, and promote eco-friendly building practices.

Sourcing Reclaimed Wood: Salvaging and Reusing Old Wood

The process of sourcing reclaimed wood involves salvaging and reusing wood from old buildings, barns, factories, and other structures that are no longer in use. This timber may come from historical sites, deconstructed buildings, or discarded wood from renovation projects.

The salvaging process begins with carefully dismantling the old structures, ensuring that the wood is preserved and salvaged without causing unnecessary damage. The reclaimed wood is then sorted and inspected for quality and usability. Any nails, screws, or other metal fasteners are removed to prepare the wood for reuse.

Once the reclaimed wood is ready, it undergoes a cleaning and restoration process, where it is sanded, planed, and treated to remove any dirt, grime, or surface imperfections. This process enhances the wood's natural beauty and prepares it for its new life in construction projects.

Utilization in Construction: Flooring

Reclaimed wood flooring has become a popular choice in

sustainable construction due to its unique character, warmth, and durability. The wide variety of wood species and patinas available in reclaimed wood flooring adds a distinctive charm to any space.

The reclaimed wood planks are carefully laid out and installed by skilled craftsmen, creating stunning floors with a rich history and story to tell. The use of reclaimed wood flooring not only reduces the demand for new timber but also diverts old wood from landfills, contributing to waste reduction efforts.

Utilization in Construction: Beams

Reclaimed wood beams are another essential application of reclaimed wood in construction. These beams, often obtained from old barns and industrial buildings, possess a weathered and aged appearance that adds a rustic and authentic touch to modern constructions.

In addition to their aesthetic appeal, reclaimed wood beams offer structural integrity and stability to building projects. They are used in various applications, including ceiling beams, support structures, and decorative elements, adding character and charm to both residential and commercial spaces.

Utilization in Construction: Decorative Elements

The versatility of reclaimed wood extends to decorative elements in construction projects. From wall cladding and accent walls to furniture and custom woodwork, reclaimed wood can be creatively incorporated to add a touch of nature and history to interior spaces.

Reclaimed wood's unique grain patterns, knots, and variations in color make it an ideal material for creating one-of-a-kind pieces that celebrate the beauty of imperfection. Whether used as shelving, tabletops, or art installations, reclaimed wood brings a sense of sustainability and authenticity to interior designs.

Environmental Benefits of Reclaimed Wood in Construction

The use of reclaimed wood in construction offers numerous environmental benefits. By salvaging and reusing old wood, construction projects contribute to waste reduction and reduce the pressure on forests and natural habitats caused by timber harvesting.

Choosing reclaimed wood also reduces the need for energy-intensive processes involved in producing new timber, such as logging, transportation, and milling. This leads to a decrease in greenhouse gas emissions and a lower carbon footprint associated with construction activities.

Using reclaimed wood promotes the preservation of historical and cultural heritage, as old structures are deconstructed with care to salvage valuable wood. By incorporating this wood into new construction projects, its story and legacy continue to be honored and celebrated.

2.9 Sheep's Wool: Natural Insulation and Renewable Resource

Sheep's wool has emerged as a sustainable and eco-friendly material for insulation in the construction industry. With its excellent insulating properties, renewable nature, and minimal environmental impact, sheep's wool insulation is gaining popularity as a viable alternative to traditional synthetic insulators.

Sourcing Sheep's Wool: Sustainable Sheep Farming

Sheep's wool is sourced from sheep, which are domesticated animals raised on farms for their wool and meat. Sustainable sheep farming practices play a crucial role in ensuring the environmental friendliness of sheep's wool as a construction material.

Sustainable sheep farming focuses on responsible animal husbandry, where sheep are provided with proper care, nutrition, and living conditions. This includes access to open pasture and grazing, reducing the need for resource-intensive feeds. Besides, sustainable sheep farmers prioritize the welfare of their animals, ensuring their health and well-being.

Shearing, the process of removing wool from sheep, is a crucial step in the sourcing of sheep's wool. Sustainable sheep farmers employ ethical shearing practices that prioritize the comfort and safety of the sheep during the process. Regular shearing helps maintain the health of the sheep and ensures the availability of high-quality wool for insulation.

Utilization in Construction: Insulation

Sheep's wool insulation offers exceptional thermal and acoustic performance, making it an ideal choice for insulating buildings. The wool's natural crimp and density create air pockets that trap heat and sound, providing effective temperature regulation and noise reduction.

During the construction process, sheep's wool insulation is applied between walls, floors, and roofs to create a comfortable and energy-efficient indoor environment. Its ability to absorb and release moisture also helps regulate humidity, contributing to improved indoor air quality.

Sheep's wool insulation is easy to work with, as it can be cut and shaped to fit various spaces and structures. Its natural fire resistance adds an additional safety feature to buildings, reducing the risk of fire spread and damage.

Utilization in Construction: Wall Coverings

Apart from its insulation properties, sheep's wool can also be utilized as a natural wall covering material. Woolen wall coverings provide a unique and visually appealing texture, adding warmth and character to interior spaces.

Sheep's wool wall coverings are available in various designs, colors, and finishes, providing flexibility in design choices for architects and interior designers. The wool's natural fibers create a soft and cozy atmosphere, enhancing the comfort and ambiance of living spaces.

Moreover, sheep's wool wall coverings contribute to improved indoor air quality as they can absorb and neutralize pollutants present in the air. This further promotes a healthy and eco-friendly indoor environment.

Environmental Benefits of Sheep's Wool in Construction

The use of sheep's wool in construction offers numerous environmental benefits. As a renewable and biodegradable resource, sheep's wool is an eco-friendly alternative to synthetic insulation materials that are derived from non-renewable fossil fuels.

The sustainable farming practices employed in sheep's wool production contribute to the conservation of natural resources and the preservation of rural landscapes. Additionally, sheep farming promotes carbon sequestration, as grass-eating sheep help maintain healthy grasslands that act as carbon sinks.

Sheep's wool insulation also reduces the demand for energy-intensive manufacturing processes associated with synthetic insulation materials. Its ability to regulate indoor temperatures reduces the need for excessive heating or cooling, leading to energy savings and reduced greenhouse gas emissions.

2.10 Clay: Traditional and Environmentally Friendly Building Material

Clay is one of the oldest and most versatile building materials used by humans throughout history. Its abundance in nature, ease of sourcing, and minimal environmental impact make it a sustainable choice for various construction applications.

Sourcing Clay: Harvesting and Processing Natural Clay Deposits

The sourcing of clay involves harvesting and processing natural clay deposits found in the earth's crust. Clay is a sedimentary material composed of fine particles, primarily composed of aluminum silicates and various minerals. These natural deposits occur in riverbeds, lakebeds, and other geological formations.

The extraction of clay typically involves surface mining or quarrying, where layers of soil and overlying materials are removed to access the clay deposits. After extraction, the clay is transported to processing facilities, where it undergoes cleaning, mixing, and refinement to achieve the desired consistency and properties for construction use.

Sustainable clay mining practices focus on minimizing the environmental impact of extraction, such as reclamation and rehabilitation of mining sites and ensuring responsible land use practices. Also, the use of efficient extraction methods helps reduce energy consumption and greenhouse gas emissions associated with clay mining operations.

Utilization in Construction: Bricks

One of the most common uses of clay in construction is in the production of bricks. Clay bricks have been widely used for thousands of years due to their durability, strength, and natural aesthetics. The process of making clay bricks involves molding clay into shape and then firing them in kilns at high temperatures to achieve hardness and stability.

Clay bricks offer excellent thermal performance, helping to regulate indoor temperatures and reducing the need for additional heating or cooling. Their natural composition also makes them fire-resistant, providing added safety to buildings.

In modern construction, clay bricks are utilized in various structural and non-structural applications, including walls, facades, and paving. They contribute to the aesthetic appeal of buildings and provide a sense of cultural heritage, especially in regions where traditional brick-making techniques are still prevalent.

Utilization in Construction: Tiles

Clay tiles are another popular application of this versatile material in construction. Clay tiles are used for roofing and flooring, offering a durable and aesthetically pleasing solution for both residential and commercial buildings.

Roofing clay tiles are designed to withstand harsh weather conditions and provide effective protection against rain, wind, and extreme temperatures. The use of clay roofing tiles also allows for proper ventilation, preventing the buildup of moisture and mold in the roof space.

Clay floor tiles, on the other hand, add an elegant touch to interior spaces. They are available in various shapes, sizes, and colors, allowing for creative and customized designs. Clay tiles are known for their resilience and longevity, making them a sustainable choice for flooring that can withstand heavy foot traffic.

Utilization in Construction: Plasters

Clay plasters offer an eco-friendly alternative to conventional cement-based plasters. Made from natural clay mixed with sand and other natural fibers, clay plasters provide a breathable and healthy interior wall finish.

The application of clay plasters helps regulate indoor humidity levels by absorbing and releasing moisture as needed. This contributes to improved indoor air quality and prevents the growth of mold and mildew. Clay plasters also act as natural insulators, reducing heat loss and improving thermal comfort within buildings.

Besides, clay plasters are easy to work with, allowing for smooth and textured finishes that add warmth and character to interior spaces. Their natural composition makes them non-toxic and suitable for people with allergies or sensitivities to chemicals found in conventional wall finishes.

Environmental Benefits of Clay in Construction

Clay is a sustainable building material with several environmental benefits. Its abundance in nature ensures a readily available and renewable resource for construction purposes. Clay mining operations can be carried out using environmentally responsible practices, minimizing the impact on ecosystems and surrounding landscapes.

The production of clay bricks and tiles requires relatively low energy inputs compared to other construction materials, leading to reduced greenhouse gas emissions. Clay products are also durable and long-lasting, which reduces the need for frequent replacements and minimizes waste generation.

2.11 Ferrock: A Carbon-Negative Alternative to Concrete

In recent years, there has been growing concern over the environmental impact of concrete production due to its significant carbon footprint. To address this issue, researchers and engineers have been exploring innovative alternatives that offer a more sustainable solution. One such alternative is Ferrock, a groundbreaking material that not only serves as a concrete replacement but also actively sequesters carbon dioxide (CO2) during its production process.

Sourcing Ferrock: Production from Industrial By-Products

Ferrock is an eco-friendly material made by combining steel dust—a by-product of industrial processes—and other recycled materials such as silica and clay. This unique composition gives Ferrock its impressive carbon-negative properties, making it a viable option for sustainable construction.

The sourcing of Ferrock materials relies on the collection of steel dust, which is generated during industrial activities like steel manufacturing and other metal-related processes. Traditionally, steel dust has been considered a waste product and posed environmental challenges, but with Ferrock, it finds a valuable purpose in the construction industry.

Utilization in Construction: Sustainable Concrete Replacement

One of the most significant advantages of Ferrock is its potential as a concrete replacement. Traditional concrete production is responsible for a substantial amount of CO_2 emissions due to the high energy consumption and chemical reactions involved in cement production. By utilizing Ferrock instead of conventional concrete, builders can significantly reduce their project's carbon footprint.

The production process of Ferrock itself is a key factor in its carbon-negative properties. During the curing phase, Ferrock absorbs CO_2 from the atmosphere and converts it into a mineral form, effectively locking away carbon and preventing its release into the environment. This transformative process helps to offset the emissions generated during the sourcing and manufacturing of the material.

Ferrock also exhibits remarkable strength and durability, making it suitable for various construction applications. It can be used in structural elements like columns, beams, and walls, as well as in non-structural components such as tiles, pavers, and decorative elements.

Environmental Benefits of Ferrock in Construction

Ferrock's carbon-negative nature is its most notable environmental benefit. By capturing CO_2 during its production, it actively contributes to combating climate change and global warming. This feature distinguishes it from conventional building materials like concrete, which releases substantial amounts of CO_2 into the atmosphere.

In addition to its carbon-negative properties, Ferrock also provides a sustainable solution for managing steel dust waste. By repurposing this by-product and incorporating it into construction materials, Ferrock reduces the environmental burden associated with industrial waste disposal.

Moreover, the use of Ferrock in construction projects promotes a circular economy, as it relies on recycled and industrial by-products rather than new raw materials. This approach reduces the demand for natural resources, lessens the strain on ecosystems, and minimizes waste generation.

Challenges and Future Prospects

While Ferrock shows great promise as a carbon-negative alternative to concrete, there are still challenges to address for its widespread adoption. One concern is the scalability of its production. As demand increases, ensuring a stable and sufficient supply of steel dust and other necessary components becomes crucial.

Additionally, more research is needed to optimize the manufacturing process and further enhance the material's properties. This includes exploring various mix designs and curing methods to achieve the desired strength, durability, and workability for different construction applications.

Ferrock represents a significant advancement in sustainable construction materials, providing a carbon-negative alternative to traditional concrete. By capturing and sequestering CO_2 during its production, Ferrock not only reduces the carbon footprint of construction projects but also actively contributes to climate change mitigation. As the construction industry seeks more environmentally friendly solutions, Ferrock emerges as a promising option for building a greener and more sustainable future.

CHAPTER 3: SUSTAINABLE CONSTRUCTION TECHNIQUES

3.1 Designing for Sustainability: Principles and Considerations

D esigning for sustainability is a fundamental aspect of creating a greener and more environmentally conscious built environment. It involves integrating principles and considerations that prioritize resource efficiency, environmental responsibility, and social equity throughout the entire lifecycle of a building. By adopting sustainable design practices, architects and engineers can minimize the environmental impact of construction projects while maximizing their positive contributions to the well-being of communities and ecosystems.

Energy efficiency is a key principle of sustainable design. It focuses on optimizing energy consumption throughout the building's life cycle, from construction to operation. By incorporating energy-efficient technologies, such as LED lighting, high-performance insulation, and smart building systems, designers can significantly reduce the building's carbon footprint and decrease its overall energy demands.

Another crucial aspect of sustainable design is the use of renewable energy sources. Integrating solar panels, wind turbines, or geothermal systems allows buildings to generate their own clean energy and reduce reliance on fossil fuels. This not only reduces greenhouse gas emissions but also lowers energy costs over time.

Water efficiency is also a significant consideration in sustainable design. Implementing water-saving fixtures, rainwater harvesting systems, and low-impact landscaping helps conserve this precious resource. Additionally, designers can explore innovative techniques like greywater recycling to further minimize water consumption.

Material selection plays a vital role in sustainable design. Opting for eco-friendly and locally sourced materials, such as bamboo, hempcrete, or reclaimed wood, reduces the environmental impact of construction and supports local economies. Also, choosing materials with low embodied energy and high recyclability contributes to a circular economy and reduces waste generation.

Designing for adaptability and flexibility ensures that buildings can meet the changing needs of occupants over time. By incorporating flexible floor plans and modular construction techniques, structures can easily adapt to future uses or technological advancements, extending their lifespan and reducing the need for demolition and reconstruction.

Biophilic design is a concept that seeks to reconnect occupants with nature by integrating natural elements into the built environment. Incorporating green spaces, natural light, and natural ventilation not only improves the health and well-being of occupants but also enhances the overall sustainability of the building.

Sustainable site planning is essential for minimizing the impact of construction on the environment. It involves careful consideration of factors such as site orientation, landscaping, and stormwater management to ensure that the building harmonizes with its surroundings and mitigates any negative effects on local ecosystems.

To achieve successful sustainable design, collaboration among architects, engineers, builders, and stakeholders is crucial. Integrating diverse perspectives and expertise from the early stages of the design process allows for innovative solutions that meet both environmental and social goals.

Considerations for Sustainable Design

Beyond the principles, specific considerations are essential for designing sustainably. These include:

Life Cycle Assessment (LCA): Evaluating the environmental impact of a building throughout its entire life cycle, from raw material extraction to demolition, helps identify opportunities for improvement and informs material and design choices.

Passive Design Strategies: Passive design focuses on maximizing natural resources, such as sunlight and ventilation, to reduce the building's reliance on artificial energy sources. Proper building orientation, shading devices, and natural ventilation systems are examples of passive design strategies.

Indoor Air Quality (IAQ): Ensuring good IAQ through proper ventilation and the use of low-emission materials promotes the health and well-being of building occupants while reducing the environmental impact of construction.

Social Equity and Inclusivity: Sustainable design should also address social issues, such as accessibility and inclusivity. Creating spaces that are welcoming and accessible to all individuals contributes to a more sustainable and equitable society.

Local Context: Designing with consideration for the local climate, culture, and context fosters a sense of place and ensures that buildings harmonize with their surroundings, reducing their ecological footprint.

Longevity and Durability: Choosing durable materials and construction methods increases the building's lifespan and reduces the need for frequent repairs and replacements, ultimately minimizing waste generation.

Post-Occupancy Evaluation (POE): Conducting POEs allows designers to assess the building's performance in real-world conditions, identify areas for improvement, and inform future design decisions.

Sustainable design is a holistic approach that considers environmental, social, and economic aspects to create buildings that are not only environmentally responsible but also contribute positively to the well-being of occupants and communities. By integrating energy-efficient technologies, renewable energy sources, and eco-friendly materials, sustainable design reduces the environmental impact of construction while promoting a healthier and more resilient built environment for future generations.

3.2 Energy-Efficient Building Design and Techniques

Energy-efficient building design and techniques are essential components of sustainable construction, aimed at reducing energy consumption and minimizing the environmental impact of buildings. By incorporating innovative technologies, thoughtful design strategies, and efficient systems, energy-efficient buildings can significantly lower greenhouse gas emissions, conserve natural resources, and provide occupants with healthier and more comfortable living and working spaces.

One of the primary considerations in energy-efficient building design is optimizing the building's envelope. The envelope, consisting of the walls, roof, windows, and doors, acts as a barrier between the interior and exterior environments. By using high-quality insulation materials, advanced glazing, and airtight construction, designers can create a well-insulated envelope that reduces heat transfer, minimizing the need for heating and cooling systems.

Passive solar design is another key technique used in energy-efficient buildings. It harnesses the natural energy from the sun to passively heat and cool the building. South-facing windows, thermal mass materials, and shading devices are strategically incorporated to optimize solar gain in winter and minimize it in summer, maximizing energy efficiency.

Energy-efficient lighting plays a vital role in reducing electricity consumption. LED lighting, in particular, is a popular choice due to its energy efficiency, long lifespan, and ability to create various lighting effects. Additionally, incorporating daylighting strategies, such as skylights and light shelves, further reduces the need for artificial lighting during daylight hours.

Heating, ventilation, and air conditioning (HVAC) systems are significant energy consumers in buildings. Employing energy-efficient HVAC systems, such as heat pumps, variable refrigerant flow systems, and demand-controlled ventilation, ensures optimal thermal comfort while reducing energy usage.

Renewable energy integration is a key aspect of energy-efficient building design. By incorporating solar panels, wind turbines, or geothermal systems, buildings can generate their own clean energy on-site, reducing reliance on grid electricity and lowering greenhouse gas emissions.

Smart building technologies, including energy management systems and smart thermostats, enable real-time monitoring and control of energy consumption. These systems optimize energy use based on occupancy patterns, weather conditions, and time of day, ensuring that energy is used efficiently and not wasted.

Building materials also play a crucial role in energy-efficient design. Choosing materials with low embodied energy, such as recycled steel and sustainably sourced wood, reduces the environmental impact of construction. Besides, using materials with high thermal resistance, such as insulated concrete forms (ICFs) and structural insulated panels (SIPs), improves the building's energy performance.

Incorporating green roofs and living walls is another technique used to enhance energy efficiency. Green roofs provide additional insulation, reducing heat gain and loss through the roof, while living walls help regulate indoor temperatures and improve indoor air quality.

Energy-efficient building design is not limited to new construction; retrofitting existing buildings is also an important strategy for reducing energy consumption. By upgrading insulation, replacing windows, and retrofitting HVAC systems, older buildings can be transformed into more energy-efficient and sustainable structures.

The benefits of energy-efficient building design extend beyond environmental conservation. They also include lower energy costs for occupants, improved indoor air quality, increased occupant comfort and productivity, and enhanced property value.

3.3 Water Conservation and Management Strategies

Water conservation and management strategies are essential components of sustainable construction, aiming to reduce water consumption, minimize water wastage, and protect this valuable natural resource. As global water scarcity becomes an increasingly pressing issue, implementing these strategies is crucial for promoting environmental sustainability and ensuring a reliable water supply for future generations.

One of the primary approaches to water conservation in buildings is the use of water-efficient fixtures and appliances. Low-flow toilets, faucets, and showerheads significantly reduce water usage without compromising performance. Additionally, energy-efficient dishwashers and washing machines further contribute to water conservation.

Rainwater harvesting is a sustainable technique that captures and stores rainwater for various uses within the building. By collecting rainwater from rooftops and directing it to storage tanks, this water can be utilized for irrigation, flushing toilets, and other non-potable applications, reducing the demand on municipal water sources.

Graywater recycling is another water conservation strategy that involves treating and reusing water from sinks, showers, and laundry for non-potable purposes. By diverting and treating graywater on-site, buildings can offset the need for fresh water in non-drinking applications, conserving potable water for essential uses.

Landscaping practices also play a significant role in water conservation. Xeriscaping, a landscaping technique that uses drought-resistant plants and efficient irrigation methods, reduces outdoor water consumption significantly. Additionally, smart irrigation systems that adjust watering schedules based on weather conditions and soil moisture levels ensure that landscapes receive just the right amount of water.

Water-efficient design extends beyond individual buildings to include community planning and development. Implementing green infrastructure solutions, such as permeable pavements and bioswales, helps manage stormwater runoff and reduce the burden on municipal drainage systems.

Building design can also incorporate features that encourage water conservation. For instance, placing water fixtures closer to hot water sources reduces the time taken for hot water to reach the fixture, minimizing water wastage. Installing dual plumbing systems for potable and non-potable water ensures that water of suitable quality is used for specific purposes.

Water metering and monitoring systems are valuable tools for managing water consumption. By tracking water usage data in real-time, building managers can identify and address leaks or inefficiencies promptly, leading to more effective water management.

Education and awareness programs are essential in promoting water conservation among building occupants. By educating residents and users about the importance of water conservation and providing tips for reducing water usage, sustainable water practices become a collective effort.

Water conservation and management strategies go beyond the building's operational phase. During construction, measures can be taken to minimize water use, such as using dust control methods to reduce water required for site stabilization and dust suppression.

3.4 Waste Reduction and Recycling in Construction

Waste reduction and recycling in construction are vital components of sustainable building practices that aim to minimize the environmental impact of the construction industry. Construction and demolition activities generate a significant amount of waste, contributing to landfills and depleting natural resources. By adopting waste reduction and recycling strategies, the construction sector can not only reduce its ecological footprint but also conserve valuable resources and promote a circular economy.

One of the fundamental principles of waste reduction in construction is proper planning and design. Adopting prefabrication and modular construction techniques can optimize material usage and minimize construction waste. By precisely calculating the required materials and ordering them in appropriate quantities, excess waste can be avoided.

Construction companies can also implement lean construction practices, which focus on efficiency and resource optimization. Lean construction aims to eliminate wasteful activities, reduce overproduction, and streamline processes, ultimately leading to reduced waste generation.

Recycling and reusing construction waste materials is a key aspect of waste reduction. Many construction materials, such as concrete, metal, wood, and asphalt, can be recycled and repurposed for future construction projects. Recycling not only diverts waste from landfills but also reduces the demand for new raw materials, conserving natural resources.

To facilitate recycling, construction companies can establish on-site recycling facilities or collaborate with recycling centers to ensure that waste materials are properly sorted and processed. This approach encourages a closed-loop system where construction waste is treated as a valuable resource.

Designing buildings with deconstruction and future recycling in mind is known as "design for disassembly." By considering how building components can be easily dismantled and recycled at the end of their life cycle, the construction industry can significantly reduce waste and make recycling more feasible.

Adopting sustainable building materials that have a lower environmental impact and can be recycled or reused is another effective waste reduction strategy. Materials such as recycled steel, reclaimed wood, and eco-friendly composites can replace traditional materials, reducing waste generation and promoting sustainable construction practices.

Waste reduction and recycling efforts are not limited to construction sites but extend to demolition activities as well. Demolition contractors can salvage and reclaim materials from old structures before demolishing them, diverting a significant amount of waste from landfills.

Government regulations and incentives can also play a crucial role in promoting waste reduction and recycling in the construction industry. Implementing policies that encourage recycling, providing financial incentives for using sustainable materials, and setting waste reduction targets can drive industry-wide adoption of sustainable practices.

Educating construction workers and stakeholders about the importance of waste reduction and recycling is essential for creating a culture of sustainability within the industry. Training programs can raise awareness about proper waste management practices, sorting techniques, and the benefits of recycling.

Waste reduction and recycling are essential components of sustainable construction practices. By incorporating proper planning and design, lean construction principles, recycling and reusing materials, and adopting sustainable building materials, the construction industry can minimize waste generation and promote a circular economy. Collaboration between construction companies, recycling centers, and government entities, along with education and training programs, are crucial for driving widespread adoption of waste reduction and recycling practices in construction. Embracing these strategies not only benefits the environment by conserving resources and reducing landfill waste but also contributes to a more sustainable and resilient construction industry.

3.5 Indoor Environmental Quality

and Health Considerations

Indoor environmental quality (IEQ) and health considerations are paramount in sustainable construction as they directly impact the well-being and productivity of building occupants. IEQ encompasses various factors that influence indoor air quality, thermal comfort, acoustics, lighting, and access to natural elements. Prioritizing these aspects in building design and construction contributes to healthier indoor environments, reduces the risk of health issues, and enhances overall occupant satisfaction and performance.

One of the critical aspects of IEQ is indoor air quality. Poor indoor air quality can result from the presence of pollutants such as volatile organic compounds (VOCs), formaldehyde, and particulate matter. These pollutants can originate from construction materials, furniture, cleaning products, and inadequate ventilation. Sustainable construction focuses on using low-VOC and non-toxic materials, ensuring proper ventilation systems, and employing air purification technologies to improve indoor air quality.

Thermal comfort is another crucial element of IEQ. Proper temperature and humidity levels play a vital role in ensuring occupant comfort and well-being. Sustainable buildings use energy-efficient heating, ventilation, and air conditioning (HVAC) systems, as well as thermal insulation, to maintain consistent indoor temperatures and reduce energy consumption.

Acoustic comfort is often overlooked but plays a significant role in occupant satisfaction. Sustainable building design incorporates sound-absorbing materials and sound insulation to minimize noise pollution from both internal and external sources, creating quieter and more peaceful indoor environments.

Lighting is an integral part of IEQ, with access to natural light being particularly beneficial. Sustainable buildings maximize the use of natural light through well-designed windows, skylights, and light shelves. This not only reduces the need for artificial lighting but also enhances occupants' well-being, mood, and productivity.

Incorporating biophilic design principles is another aspect of IEQ that focuses on connecting building occupants with nature. Providing views of green spaces, integrating indoor plants, and incorporating natural elements into interior design contribute to a sense of well-being and improve mental health.

Health considerations in sustainable construction extend beyond IEQ. Building materials and finishes play a vital role in ensuring a healthy indoor environment. Choosing non-toxic, allergen-free, and mold-resistant materials helps prevent health issues such as allergies and respiratory problems.

Moreover, sustainable construction prioritizes the use of renewable and eco-friendly materials that do not emit harmful substances over their life cycle. For example, opting for sustainably sourced wood and eco-friendly paints can significantly contribute to a healthier indoor environment.

In addition to the direct health benefits, sustainable construction also supports the overall well-being of communities by incorporating social aspects. Creating spaces that encourage physical activity, promote social interactions, and support mental well-being enhances the quality of life for building occupants.

Certification systems such as LEED (Leadership in Energy and Environmental Design) and WELL Building Standard provide guidelines and criteria for achieving high levels of IEQ and health considerations in buildings. By striving for these certifications, construction projects can ensure they meet rigorous standards for environmental sustainability and occupant well-being.

3.6 Sustainable Site Selection and Planning
Sustainable site selection and planning are critical aspects of

sustainable construction that aim to minimize the environmental impact of a building project and promote responsible land use. By carefully considering the site and its surrounding environment, construction projects can make a positive contribution to the ecosystem and the community while reducing resource consumption and energy use.

One key principle of sustainable site selection is to prioritize brownfield redevelopment over greenfield development. Brownfield sites are previously developed areas that may be underutilized or contaminated. Choosing to redevelop these sites helps revitalize urban areas, reduce urban sprawl, and preserve natural habitats and open spaces.

In contrast, greenfield development involves building on previously undeveloped land, often leading to the loss of valuable ecosystems and farmland. By avoiding greenfield development and focusing on brownfield sites, sustainable construction projects can make a significant difference in preserving natural resources and protecting biodiversity.

Sustainable site planning involves conducting thorough site assessments to identify any environmental sensitivities and natural features that should be protected. This may include wetlands, water bodies, native vegetation, and wildlife habitats. By preserving and incorporating these elements into the design, construction projects can support biodiversity and promote ecological balance.

Another important consideration in sustainable site planning is reducing the building's environmental footprint. This includes minimizing site disturbance during construction, preserving natural drainage patterns, and using permeable surfaces to reduce stormwater runoff and promote groundwater recharge.

Sustainable site planning also involves considering transportation and accessibility. Locating projects near public transportation hubs and providing amenities for walking and biking can reduce the need for private vehicle use and lower greenhouse gas emissions. Additionally, designing pedestrian-friendly spaces enhances community connectivity and encourages a healthier lifestyle.

Implementing sustainable site planning often involves collaboration with local communities and stakeholders. Engaging with the community allows construction projects to address concerns and incorporate input from residents, ensuring that the development aligns with the needs and values of the people who will live and work in the area.

Sustainable site planning also takes into account the long-term impacts of the construction project. This includes considering climate change and adapting the design to withstand potential future challenges, such as extreme weather events and rising sea levels.

Incorporating green infrastructure is another essential aspect of sustainable site planning. Green infrastructure refers to the use of natural systems and features, such as green roofs, rain gardens, and bioswales, to manage stormwater and provide ecological benefits. Green infrastructure can help improve water quality, reduce the urban heat island effect, and support biodiversity.

Finally, sustainable site selection and planning also consider the project's proximity to essential services and amenities, such as schools, healthcare facilities, and recreational areas. Ensuring easy access to these services contributes to the well-being and convenience of building occupants and fosters a sense of community.

3.7 Green Building Certifications and Standards

Green building certifications and standards play a vital role in promoting sustainable construction practices and guiding the development of environmentally friendly buildings. These certifications and standards are established by various organizations and government agencies to set benchmarks and best practices for sustainable design, construction, operation, and maintenance of buildings. They provide a clear framework for evaluating a building's environmental performance and help create healthier, more resource-efficient, and environmentally responsible structures.

One of the most well-known green building certifications is Leadership in Energy and Environmental Design (LEED), developed and maintained by the U.S. Green Building Council (USGBC). LEED provides a rating system that evaluates building projects across several categories, including site selection, water efficiency, energy and atmosphere, materials and resources, indoor environmental quality, and innovation in design. The rating system assigns points for meeting specific criteria within each category, and buildings can achieve different levels of LEED certification, such as Certified, Silver, Gold, or Platinum, based on the number of points earned.

Another prominent green building certification is the BREEAM (Building Research Establishment Environmental Assessment Method) certification, which originated in the United Kingdom but is now used worldwide. BREEAM assesses buildings based on a range of environmental and sustainability factors, including energy and water use, materials selection, waste management, and ecological impact. Like LEED, BREEAM offers different certification levels, from Pass to Outstanding, depending on the building's overall environmental performance.

The Green Building Initiative (GBI) administers the Green Globes certification, which is an interactive online assessment tool that provides guidance and certification for green building projects. Green Globes evaluates buildings on various aspects, such as energy efficiency, water conservation, indoor air quality, and environmental impact. It offers a flexible and customizable approach, allowing projects to choose the elements that align best with their sustainability goals.

In addition to these certifications, many countries have their own national green building standards. For example, Australia has the Green Star rating system, Germany has the DGNB (Deutsche Gesellschaft für Nachhaltiges Bauen) certification, and China has the Three-Star System for evaluating the environmental performance of buildings.

Green building certifications and standards drive the adoption of sustainable practices in the construction industry by providing clear guidelines, promoting innovation, and rewarding environmentally responsible projects. They encourage developers, architects, and builders to prioritize sustainability and consider the long-term impacts of their projects on the environment, occupants, and the community.

By obtaining green building certifications, construction projects can demonstrate their commitment to sustainability, gain recognition for their efforts, and attract environmentally conscious investors, tenants, and customers. Additionally, green building certifications often lead to reduced operating costs through improved energy and water efficiency, which can result in significant savings over the life of the building.

These certifications also foster a culture of continuous improvement, encouraging building owners to monitor and optimize their building's performance throughout its life cycle. They often require ongoing data collection and reporting, which helps identify opportunities for further enhancements and encourages the adoption of innovative technologies and practices.

Green building certifications are not only applicable to new construction but can also be applied to existing buildings through retrofitting and renovation projects. Retrofitting existing buildings to meet green building standards can significantly improve their energy efficiency, reduce their environmental impact, and extend their service life.

CHAPTER 4: SERVICE-LIFE PREDICTION AND MAINTENANCE

4.1 Assessing the Lifespan of Sustainable Construction Materials

Assessing the lifespan of sustainable construction materials is a crucial aspect of ensuring the long-term viability and environmental impact of building projects. Sustainable construction materials are chosen not only for their eco-friendly attributes but also for their durability and ability to withstand the test of time. In this section, we will delve into the factors that influence the lifespan of these materials and the methods used to evaluate their performance over time.

One of the key considerations when assessing the lifespan of sustainable construction materials is their inherent durability and resistance to degradation. Unlike traditional materials, such as concrete or steel, which may have a limited lifespan due to corrosion or deterioration, sustainable materials are often chosen for their ability to withstand environmental stresses. For example, bamboo, a renewable and versatile material, possesses natural durability and can be engineered to be as strong as hardwood. This makes it suitable for a wide range of construction applications, from structural elements to flooring and wall panels.

Similarly, hempcrete, made from hemp shives and a lime-based binder, exhibits excellent insulation properties and breathability, leading to improved longevity compared to conventional building materials. Hempcrete structures have been found to remain stable and structurally sound for many decades, reducing the need for frequent repairs or replacements.

In the case of cork, another sustainable material known for its thermal and acoustic insulation properties, its cellular structure contributes to its resilience and resistance to wear and tear. When used as flooring or wall coverings, cork can maintain its integrity and aesthetic appeal for years.

To accurately assess the lifespan of sustainable construction materials, various evaluation methods and performance metrics are employed. Accelerated aging tests and exposure to extreme environmental conditions allow researchers and engineers to simulate the effects of decades of real-world use in a compressed timeframe. These tests help identify any weaknesses or potential issues that may arise over time, enabling designers to refine the material's composition and enhance its durability.

Another crucial aspect of assessing lifespan is the material's ability to adapt to changing environmental conditions and climate impacts. Sustainable materials often exhibit better resilience to weather extremes, such as temperature fluctuations, moisture, and UV radiation. As climate change becomes an increasingly pressing issue, building materials that can endure and perform well under evolving conditions will be essential for ensuring the longevity of structures.

Additionally, the manner in which sustainable materials are incorporated into the construction process can influence their lifespan. Proper installation, maintenance, and adherence to manufacturer guidelines are essential to maximize the materials' longevity. For instance, proper curing and sealing of hempcrete ensure that it reaches its full strength and durability potential.

Monitoring and regular inspections throughout a building's life cycle also help identify signs of wear, damage, or deterioration. Implementing timely repairs and maintenance can significantly extend the lifespan of sustainable materials, reducing the need for replacements and minimizing overall environmental impact.

Assessing the lifespan of sustainable construction materials is a multidimensional process that considers inherent durability, performance under various environmental conditions, and proper installation and maintenance. The continuous refinement and improvement of these materials, along with comprehensive monitoring and inspections, will contribute to more resilient and long-lasting structures that align with the principles of sustainability. By incorporating materials with extended lifespans, the construction industry can contribute to a greener and more environmentally conscious future.

4.2 Predictive Maintenance Strategies for Sustainable Buildings

Predictive maintenance strategies play a crucial role in ensuring the long-term performance and sustainability of buildings. Unlike traditional reactive maintenance, which involves fixing issues only after they arise, predictive maintenance takes a proactive approach by using data and technology to anticipate potential problems before they become major failures. In sustainable buildings, implementing predictive maintenance strategies can lead to significant cost savings, reduced environmental impact, and improved occupant comfort and safety.

One of the key benefits of predictive maintenance in sustainable buildings is its ability to optimize the use of resources. By continuously monitoring building systems, such as HVAC (heating, ventilation, and air conditioning) and lighting, predictive maintenance can identify inefficiencies and potential malfunctions. This allows building managers to address these issues promptly, preventing energy waste and reducing utility costs.

For instance, an intelligent building management system equipped with sensors and data analytics can analyze energy consumption patterns and detect anomalies. If it identifies a gradual increase in energy usage in a specific area, it may indicate a malfunctioning HVAC unit. By addressing the problem early, building operators can avoid excessive energy consumption and ensure optimal system performance.

Predictive maintenance also plays a vital role in maximizing the lifespan of sustainable construction materials. As mentioned in the previous section, sustainable materials are selected for their durability and resilience. However, environmental factors and wear over time can still impact their performance. By monitoring the condition of these materials through sensors and inspection data, predictive maintenance can help identify signs of degradation and deterioration. Timely repairs or maintenance can then be scheduled to extend the materials' lifespan and prevent unnecessary replacements.

Occupant comfort and well-being are essential aspects of sustainable building design. Predictive maintenance can contribute to improved occupant satisfaction by ensuring that building systems, such as indoor air quality, temperature, and lighting, remain at optimal levels. Indoor air quality, in particular, is crucial for occupant health and productivity. Regular monitoring and analysis of air quality data can help detect potential issues, such as high levels of pollutants or inadequate ventilation, and allow for timely remediation.

Predictive maintenance strategies can also enhance occupant safety by identifying potential hazards and risks. For example, sensors can monitor structural integrity, detecting any signs of structural fatigue or damage. This early warning system can prompt building managers to take appropriate measures, ensuring the safety of occupants.

Another aspect where predictive maintenance proves beneficial is in reducing downtime and disruptions. In sustainable buildings, where energy-efficient and renewable technologies are often integrated, sudden system failures can have a more significant impact on building operations. Predictive maintenance can prevent such disruptions by providing real-time insights into equipment performance, allowing for proactive maintenance or replacement.

The implementation of predictive maintenance strategies in sustainable buildings is facilitated by advancements in building automation, the Internet of Things (IoT), and data analytics. Smart building technologies continuously collect and analyze data from various sensors and devices, enabling real-time monitoring and decision-making. The use of artificial intelligence and machine learning algorithms further enhances the predictive capabilities of these systems, as they learn from historical data to identify patterns and trends.

4.3 Extending the Lifespan of Eco-Friendly Structures

Extending the lifespan of eco-friendly structures is essential for maximizing the environmental and economic benefits of sustainable construction. While sustainable buildings are designed to be durable and resilient, proper maintenance and periodic upgrades are necessary to ensure their longevity. Several strategies can be employed to extend the lifespan of eco-friendly structures, ranging from routine maintenance and repairs to adaptive reuse and retrofitting.

Regular maintenance is a fundamental aspect of preserving any building, and eco-friendly structures are no exception. Building owners and facility managers should establish a comprehensive maintenance plan that includes routine inspections, cleaning, and minor repairs. Regular inspections can help identify potential issues early on, allowing for prompt repairs and preventing small problems from escalating into major failures. This proactive approach not only extends the lifespan of building components but also helps to maintain optimal energy efficiency and occupant comfort.

In addition to routine maintenance, sustainable buildings can benefit from specialized maintenance practices that cater to the unique characteristics of eco-friendly materials. For example, for structures with green roofs or living walls, proper maintenance of the vegetation and drainage systems is crucial to ensure their functionality and prevent water damage. For buildings with solar panels, regular cleaning and inspection are essential to maximize energy generation.

Utilizing durable and high-quality eco-friendly materials during construction is a key factor in extending the lifespan of sustainable structures. Careful consideration should be given to selecting materials that are resistant to environmental factors, such as moisture, temperature fluctuations, and UV radiation. For instance, using weather-resistant cladding materials and protective coatings can help prevent water infiltration and degradation over time.

Implementing adaptive reuse strategies can significantly extend the lifespan of eco-friendly structures while minimizing waste and environmental impact. Adaptive reuse involves repurposing existing buildings for new functions, thereby giving them a new lease on life. Rather than demolishing and constructing new buildings, adaptive reuse allows for the preservation of valuable resources and the retention of embodied energy in the existing structure. This approach aligns with the principles of sustainability and circular economy, where resources are kept in use for as long as possible.

Retrofitting existing eco-friendly structures with innovative technologies and systems can enhance their performance and extend their lifespan. As sustainable technologies evolve, older buildings may benefit from upgrades that improve energy efficiency, indoor air quality, and overall comfort. For example, retrofitting a building with more energy-efficient windows, advanced insulation, or smart building systems can lead to significant energy savings and a longer service life.

Collaboration among architects, engineers, and construction professionals is essential in designing eco-friendly structures with longevity in mind. Considering the building's lifecycle and designing for adaptability and flexibility can future-proof the structure and accommodate changing needs and technologies. Additionally, creating detailed as-built documentation and maintaining accurate records of construction and material specifications can assist in future maintenance and renovation efforts.

Education and awareness among building occupants and users also play a role in extending the lifespan of eco-friendly structures. Teaching occupants about proper energy usage, waste management, and sustainable practices can help minimize wear and tear on building systems and materials. Additionally, engaging occupants in sustainability initiatives fosters a sense of responsibility and stewardship towards the building, leading to better care and preservation.

CHAPTER 5:
CONSTRUCTION 4.0
AND DIGITALIZATION

5.1 The Role of Technology in Sustainable Construction

Technology plays a crucial role in advancing sustainable construction practices and driving positive environmental outcomes in the building industry. With the increasing global focus on environmental conservation and reducing the carbon footprint, integrating technology into sustainable construction has become more vital than ever. From design and planning to construction and operation, innovative technologies offer a wide array of tools and solutions that enhance efficiency, minimize resource consumption, and promote eco-friendly building practices.

One of the significant contributions of technology to sustainable construction is Building Information Modeling (BIM). BIM is a digital representation of the physical and functional characteristics of a building, providing a collaborative platform for architects, engineers, contractors, and other stakeholders to work together seamlessly. BIM allows for efficient project planning, clash detection, and visualization, reducing material waste and energy consumption during the construction phase. Additionally, BIM facilitates the integration of sustainable design principles, enabling professionals to optimize energy performance, daylighting, and HVAC systems for improved building efficiency.

Artificial Intelligence (AI) and machine learning have also found their way into sustainable construction. AI algorithms can analyze vast datasets to identify patterns and optimize building designs for energy efficiency. They can predict energy consumption and performance, helping architects and engineers make informed decisions about building materials and systems. AI-powered sensors and smart building technologies enable real-time monitoring of energy usage, indoor air quality, and temperature, allowing for proactive maintenance and energy-saving strategies.

The advent of the Internet of Things (IoT) has revolutionized sustainable building management. IoT-connected devices and sensors enable the creation of smart buildings that can self-regulate energy usage and optimize resource allocation. For instance, smart lighting systems can adjust brightness based on occupancy, while automated climate control systems can adapt heating and cooling based on real-time occupancy and weather conditions. By reducing energy waste and improving occupant comfort, IoT technologies contribute to overall building sustainability.

Renewable energy technologies have gained significant traction in sustainable construction. Solar panels, wind turbines, and geothermal systems provide clean and renewable sources of energy, reducing buildings' dependence on fossil fuels and lowering greenhouse gas emissions. Integrating renewable energy systems into the building's design allows for on-site power generation, enabling sustainable buildings to become energy self-sufficient or even contribute excess energy back to the grid.

Prefabrication and modular construction are technology-driven methodologies that promote sustainability in the construction process. Prefabricated building components are manufactured off-site, reducing construction waste and improving construction efficiency. Modular construction techniques allow for more precise and controlled construction, resulting in higher-quality buildings with minimized material waste.

Virtual Reality (VR) and Augmented Reality (AR) are transforming the way buildings are designed and experienced. VR allows architects and clients to virtually walk through a building before construction begins, giving them a realistic sense of space and design. AR overlays digital information onto physical spaces, facilitating on-site construction and maintenance tasks. By streamlining the design and construction process, VR and AR technologies contribute to better planning, reduced errors, and overall project sustainability.

Big Data analytics is yet another technology that has found applications in sustainable construction. By collecting and analyzing vast amounts of data, construction professionals can identify areas of improvement, optimize resource usage, and make data-driven decisions that lead to more sustainable outcomes. Big Data can also be used to track and measure a building's environmental performance, helping assess its energy efficiency and identifying areas for improvement.

As the demand for sustainable buildings continues to grow, technology will remain at the forefront of sustainable construction, facilitating the creation of greener, smarter, and more resilient structures.

5.2 Building Information Modeling (BIM) for Eco-Friendly Design

Building Information Modeling (BIM) is a powerful tool that has become a cornerstone of eco-friendly design and sustainable construction practices. BIM is a collaborative, 3D digital modeling process that encompasses the entire building lifecycle, from conceptualization and design to construction, operation, and eventual demolition or renovation. By providing a comprehensive and integrated platform for architects, engineers, contractors, and other stakeholders to work together, BIM optimizes building efficiency, minimizes waste, and enhances environmental performance.

One of the key advantages of BIM in eco-friendly design is its ability to facilitate sustainable decision-making during the early stages of a project. BIM allows designers to explore various design alternatives and evaluate their environmental impact, such as energy consumption, carbon emissions, and material usage. Through energy analysis tools, architects can assess the building's energy efficiency and identify opportunities for improvement. They can also simulate daylighting and thermal performance to optimize passive design strategies, reducing the need for artificial lighting and heating.

BIM also enables the integration of sustainable materials and systems into the building's design. By providing a centralized platform for specifying eco-friendly materials and products, BIM ensures that green building standards and certifications are met. Designers can choose materials with low environmental impact, such as recycled or rapidly renewable resources, and track their usage throughout the project. Additionally, BIM allows for the visualization of the entire supply chain, promoting transparency and sustainability in material sourcing and procurement.

During the construction phase, BIM continues to contribute to eco-friendly practices. The detailed 3D models generated by BIM aid in clash detection and coordination, reducing errors and rework, which in turn minimizes material waste and resource consumption. BIM also enhances construction scheduling and sequencing, allowing for just-in-time delivery of materials, reducing on-site storage needs and transportation-related emissions.

Once a building is operational, BIM can play a significant role in optimizing its energy performance and overall sustainability. Real-time data from sensors and smart building technologies can be integrated into BIM models, enabling continuous monitoring of energy usage, indoor air quality, and other environmental parameters. This data-driven approach allows facility managers to implement energy-saving measures and preventive maintenance strategies, maximizing the building's operational efficiency and minimizing its environmental impact.

BIM also supports sustainable facility management and maintenance throughout the building's lifecycle. As buildings age, they require regular maintenance and occasional renovations. BIM models, combined with data on building performance, can inform decision-making during retrofit projects, ensuring that sustainability objectives are maintained or improved over time. BIM also aids in the deconstruction and recycling of materials at the end of a building's life, promoting circular economy principles and reducing waste sent to landfills.

5.3 Internet of Things (IoT) Applications in Sustainable Building

The Internet of Things (IoT) has emerged as a game-changer in sustainable building design, construction, and operation. IoT refers to the network of interconnected devices embedded with sensors and software that collect and exchange data over the internet. In the context of sustainable building practices, IoT applications offer unprecedented opportunities for enhancing energy efficiency, optimizing resource usage, and improving overall environmental performance.

One of the key areas where IoT applications have a significant impact is in building automation and control systems. IoT-enabled smart devices, such as thermostats, lighting controls, and occupancy sensors, can be integrated into building management systems to regulate heating, cooling, lighting, and ventilation based on real-time data. This dynamic control allows for better energy management, as the building's systems can respond to changing conditions and occupant behavior, resulting in reduced energy consumption and lower operating costs.

IoT also plays a vital role in monitoring and optimizing water usage in sustainable buildings. Smart water meters and sensors can track water flow and detect leaks or inefficiencies, enabling timely repairs and better management of water resources. By identifying patterns of water usage and wastage, building managers can implement water conservation strategies, such as using recycled water for non-potable purposes or installing water-efficient fixtures.

In sustainable buildings, lighting constitutes a significant portion of energy consumption. IoT-enabled lighting systems can adjust illumination levels based on occupancy and natural lighting conditions, ensuring that lights are only on when needed. Additionally, these systems can be integrated with daylight harvesting technologies to maximize the use of natural light, further reducing energy usage and creating a more comfortable and productive indoor environment.

The application of IoT in waste management is another area of significant impact. Smart waste management systems equipped with sensors can monitor waste levels in trash bins and compactors. When waste reaches a predetermined level, the system can automatically trigger waste collection, optimizing waste pick-up routes and reducing unnecessary trips, thereby lowering fuel consumption and greenhouse gas emissions associated with waste collection.

IoT applications also contribute to enhancing occupant comfort and well-being. In smart buildings, occupants can have control over their immediate environment through smartphone apps or voice-activated systems. They can adjust temperature settings, lighting preferences, and even access real-time air quality data, empowering them to create personalized, comfortable, and healthy indoor spaces.

IoT also plays a crucial role in preventive maintenance and asset management. Sensors embedded in building equipment and systems can monitor their performance and health status, providing early warnings of potential issues. This proactive approach allows facility managers to schedule maintenance activities before equipment failures occur, extending the lifespan of assets and minimizing downtime.

To realize the full potential of IoT in sustainable building applications, data analytics and machine learning come into play. Analyzing the massive amounts of data generated by IoT devices can provide valuable insights into building performance and occupant behavior. Machine learning algorithms can identify patterns, predict energy consumption, and recommend optimal settings for energy efficiency and occupant comfort.

CHAPTER 6: CIRCULAR ECONOMY IN CONSTRUCTION

6.1 Understanding the Circular Economy Concept

The circular economy is a concept and economic model that aims to redefine the traditional linear "take, make, dispose" approach to resource consumption and waste generation. Instead of the traditional linear model, the circular economy advocates for a closed-loop system, where products, materials, and resources are continuously reused, refurbished, remanufactured, and recycled to extend their lifespan and minimize waste. The core principle of the circular economy is to create a regenerative system that reduces resource depletion, pollution, and environmental impacts while fostering sustainable economic growth.

In a circular economy, products are designed with durability, repairability, and recyclability in mind. This means that manufacturers consider the entire lifecycle of a product, from sourcing raw materials to end-of-life disposal, and design products to be easily disassembled and their components reused or repurposed. By incorporating circular design principles, products are less likely to become obsolete quickly, reducing the need for frequent replacements and minimizing waste.

The circular economy is not just limited to individual products; it also applies to entire industries and supply chains. Instead of operating in isolation, businesses collaborate and share resources, creating synergies that optimize resource use and reduce waste. For example, one company's waste or by-product can become a valuable input for another company's production process, closing the loop and creating a more circular flow of materials and resources.

One of the key advantages of the circular economy is its potential to decouple economic growth from resource consumption and environmental impacts. In a traditional linear economy, economic growth is often accompanied by increased resource extraction and waste generation, leading to environmental degradation. In contrast, the circular economy seeks to achieve prosperity while minimizing the use of finite resources and reducing environmental harm.

The circular economy also offers significant economic benefits. By keeping materials and products in circulation for longer, businesses can reduce their dependence on costly raw materials, leading to cost savings and increased competitiveness. Additionally, the circular economy fosters innovation, as businesses seek new ways to repurpose and recycle materials, leading to the development of new technologies and business models.

To transition to a circular economy, collaboration among various stakeholders is crucial. Governments, businesses, consumers, and non-governmental organizations all play a role in promoting circularity. Governments can enact supportive policies and regulations that incentivize circular practices, such as extended producer responsibility (EPR) laws that hold manufacturers responsible for the entire lifecycle of their products. Businesses can adopt circular design principles, implement recycling and remanufacturing programs, and explore new business models, such as product-as-a-service, where customers lease products instead of purchasing them outright. Consumers can also contribute by making conscious choices, such as buying products with longer lifespans and recycling or repurposing items at the end of their use.

6.2 Implementing Circular Economy Principles in Building Projects

Implementing circular economy principles in building projects involves adopting a holistic approach from the initial design stages through construction, operation, and eventual deconstruction or renovation. By embracing these principles, the construction industry can significantly reduce its environmental impact, enhance resource efficiency, and contribute to a more sustainable built environment.

One key strategy is to design buildings for disassembly and reuse. This means emphasizing modular and adaptable design that allows for easy disassembly of building components, such as walls, floors, and ceilings. By doing so, materials can be salvaged and reused during renovation or at the end of a building's life, reducing waste and preserving valuable resources.

Another important aspect is material selection and procurement. Opting for sustainable and environmentally-friendly materials is crucial in a circular economy approach. This may include using recycled or upcycled materials, as well as choosing materials with a lower environmental footprint in terms of extraction, production, and transportation.

Incorporating renewable energy sources is another vital consideration. Building projects should prioritize the integration of solar panels, wind turbines, or other renewable energy systems to reduce reliance on fossil fuels and minimize greenhouse gas emissions during the building's operational phase.

Energy efficiency measures play a significant role in circular economy practices. Implementing energy-efficient technologies, such as LED lighting, advanced insulation, and smart building management systems, can lead to reduced energy consumption and operating costs, while simultaneously lowering the building's environmental impact.

Water conservation and management are equally important aspects of sustainable construction. Implementing water-efficient fixtures and systems, as well as utilizing rainwater harvesting and greywater recycling, can significantly reduce water consumption and help preserve this precious resource.

Adopting green building certifications and standards, such as LEED (Leadership in Energy and Environmental Design) or BREEAM (Building Research Establishment Environmental Assessment Method), can provide a framework for implementing circular economy principles and ensuring sustainable building practices are met throughout the project's lifecycle.

6.3 Materials Reuse, Recycling, and Upcycling in Construction

Materials reuse, recycling, and upcycling are essential components of sustainable construction practices that align with the principles of the circular economy. These strategies aim to minimize waste generation, reduce resource consumption, and extend the lifespan of materials, ultimately contributing to a more environmentally responsible and resource-efficient building industry.

Reuse involves salvaging and repurposing materials from existing buildings or construction sites for use in new projects. This process not only reduces the demand for new materials but also prevents valuable resources from ending up in landfills. Commonly reused materials include structural elements like beams, columns, and bricks, as well as finishes such as doors, windows, and flooring. By carefully deconstructing and salvaging materials from older buildings, construction professionals can extend the life of these components and give them a new purpose, reducing the need for raw materials and minimizing environmental impact.

Recycling is another crucial aspect of sustainable construction, involving the processing of waste materials to create new products or materials. Materials such as concrete, asphalt, metals, plastics, and glass can be recycled and transformed into aggregates, road base, or even new building products. Recycling not only diverts waste from landfills but also conserves energy and resources that would have been required to produce new materials from scratch. The integration of recycling facilities and programs within construction sites is instrumental in maximizing the recovery of valuable materials and promoting a more sustainable construction industry.

Upcycling takes recycling a step further by transforming discarded materials or products into higher-value items with enhanced functionality or aesthetics. Rather than breaking materials down into raw materials for new products, upcycling involves repurposing them into more valuable components. For example, reclaimed wood from old buildings can be upcycled into decorative elements, furniture, or even artistic installations, adding unique character and value to a new construction project.

Implementing materials reuse, recycling, and upcycling in construction requires collaboration among various stakeholders, including architects, designers, contractors, and waste management companies. Designing buildings for disassembly, as discussed earlier, facilitates the process of material salvage and reuse during renovations or deconstruction.

Integrating waste separation and collection systems on construction sites also encourages responsible waste management and maximizes the potential for recycling and upcycling. Construction companies can also establish partnerships with recycling facilities and vendors specializing in reclaimed materials to ensure a steady supply of reused and upcycled components for their projects.

Government regulations and incentives play a vital role in driving the adoption of materials reuse, recycling, and upcycling in the construction industry. By incentivizing sustainable practices and setting waste reduction targets, policymakers can encourage construction companies to prioritize these strategies and contribute to a more circular and resource-efficient economy.

CHAPTER 7:
SUSTAINABLE
PROCUREMENT AND
MATERIAL TRANSPORT

7.1 Sustainable Materials Procurement Practices

Sustainable materials procurement practices are essential for promoting environmental responsibility, social equity, and economic viability in the construction industry. These practices involve sourcing and acquiring construction materials in a manner that prioritizes eco-friendly options, reduces the environmental footprint, and supports ethical supply chains. By adopting sustainable procurement practices, construction companies can play a significant role in mitigating environmental impacts and fostering a more sustainable future.

One of the key aspects of sustainable materials procurement is selecting environmentally friendly and responsibly sourced materials. This involves conducting thorough research and due diligence to identify suppliers that adhere to sustainable practices. Companies can prioritize materials with eco-label certifications, such as Forest Stewardship Council (FSC) certification for wood products or Cradle to Cradle (C2C) certification for various building materials. These certifications ensure that the materials are sourced from responsibly managed forests or are designed for circularity, promoting a closed-loop system where materials can be recycled or upcycled at the end of their life cycle.

In addition to environmental considerations, sustainable procurement practices also encompass social and ethical factors. Companies should assess suppliers' labor practices, fair wages, and commitment to workers' safety and rights. Ethical sourcing of materials ensures that workers are treated fairly and that their well-being is not compromised during the production process. By supporting suppliers with strong social responsibility practices, construction companies can contribute to positive social impacts and contribute to a more equitable supply chain.

Another crucial aspect of sustainable materials procurement is reducing the distance materials need to travel. Sourcing materials locally or regionally can significantly reduce transportation-related emissions and promote the use of locally available resources. By choosing nearby suppliers, construction companies can also support local economies and foster community engagement. Additionally, utilizing recycled or reclaimed materials can further reduce the environmental impact by diverting waste from landfills and reducing the demand for new raw materials.

To implement sustainable materials procurement practices effectively, construction companies can integrate sustainability criteria into their procurement policies and guidelines. This involves setting specific sustainability goals, establishing supplier evaluation criteria, and ensuring that sustainable practices are prioritized during the selection process. Companies can also foster collaboration with suppliers to encourage continuous improvement in sustainability performance and ensure transparency in the supply chain.

Leveraging technology and digital tools can also

enhance sustainable materials procurement practices. Building Information Modeling (BIM) and other digital platforms can assist in material tracking, supply chain management, and lifecycle analysis. By harnessing the power of data and analytics, construction companies can make informed decisions, optimize material usage, and identify opportunities for improvement.

7.2 Transporting Sustainable Construction Materials to the Site

Transporting sustainable construction materials to the construction site is a critical phase that requires careful planning and execution to minimize environmental impacts and ensure the integrity of the materials. Sustainable transportation practices focus on reducing emissions, energy consumption, and overall environmental footprint while optimizing efficiency and maintaining the quality of the materials.

One of the primary strategies for sustainable transportation is to prioritize local sourcing of materials whenever possible. By choosing suppliers located in close proximity to the construction site, the need for long-distance transportation is minimized, reducing fuel consumption and greenhouse gas emissions associated with transportation. Local sourcing also supports regional economies and communities, fostering a more sustainable and resilient supply chain.

To further reduce emissions, construction companies can explore alternative transportation methods. Transitioning to electric or hybrid vehicles for material delivery can significantly lower carbon emissions and air pollution. Alternatively, companies can utilize eco-friendly modes of transportation, such as bicycles or electric cargo bikes, for smaller-scale material deliveries within urban areas. Implementing efficient route planning and delivery scheduling can also optimize transportation and minimize unnecessary trips.

Consolidation of material shipments is another effective strategy for sustainable transportation. By combining multiple material orders into a single shipment, construction companies can reduce the number of transportation trips required, thereby lowering fuel consumption and emissions. Collaboration with other construction projects in the vicinity to consolidate orders can lead to even greater efficiency gains and cost savings.

Incorporating smart technology and real-time tracking systems can further enhance sustainable transportation practices. Internet of Things (IoT) devices and telematics can be used to monitor vehicle performance, optimize routes, and provide real-time updates on delivery status. This data-driven approach enables better decision-making, reduces idle time, and enhances overall logistics efficiency.

Another aspect to consider in sustainable transportation is the packaging of materials. Choosing eco-friendly packaging materials, such as recyclable or reusable containers, can help minimize waste and environmental impact. Additionally, companies can work with suppliers to adopt responsible packaging practices and reduce unnecessary packaging materials.

Collaboration and communication are crucial in implementing sustainable transportation practices. Construction companies should engage with suppliers and transportation partners to align on sustainability goals and jointly develop strategies for reducing environmental impacts. By fostering a culture of sustainability throughout the supply chain, all stakeholders can work together towards a common objective of reducing transportation-related emissions and promoting sustainable construction practices.

7.3 Field Storage and Installation of Sustainable Materials

Field storage and installation of sustainable materials are critical aspects of sustainable construction practices. Proper handling, storage, and installation procedures ensure that sustainable materials maintain their quality and performance, minimizing waste and maximizing their environmental benefits throughout the construction process.

When it comes to field storage of sustainable materials, several key considerations come into play. First and foremost is the need to protect materials from exposure to the elements. Many sustainable materials, such as bamboo, hempcrete, and cork, are sensitive to moisture, which can lead to mold, decay, or other forms of damage. Therefore, construction teams must store these materials in dry and covered areas to prevent water infiltration.

For instance, bamboo should be kept off the ground and stored in a cool, dry location to prevent swelling and decay. Hempcrete, which is a mixture of hemp shives and lime binder, is also susceptible to moisture, so it should be protected from rain during storage. Cork, being a natural material, can expand and contract with fluctuations in humidity, so proper climate control during storage is essential.

In addition to protecting materials from moisture, construction teams must also safeguard them from excessive heat, direct sunlight, and temperature variations. Some sustainable materials can be sensitive to temperature changes, and exposure to extreme conditions may compromise their structural integrity or aesthetic appeal. Implementing shade and insulation measures during storage can help mitigate these risks.

Adequate labeling and organization of sustainable materials in storage are also crucial. Properly identifying each material and its intended use helps prevent mix-ups and reduces the risk of using the wrong materials in construction, which could result in rework and waste.

When it comes to the installation of sustainable materials,

following manufacturer guidelines and industry best practices is essential. Different sustainable materials may have specific installation requirements, and construction teams must adhere to them to ensure optimal performance and longevity.

For example, when using bamboo as a structural element, it's essential to inspect each bamboo piece for defects and ensure proper connections and bracing for stability. For hempcrete installation, proper mixing of hemp shives and lime binder, along with accurate layering and compaction, is critical to achieving the desired insulation and structural properties.

During installation, construction teams should also aim to minimize waste generation by carefully measuring and cutting materials to fit the required dimensions. Proper training and skill development among construction workers are essential to ensure that sustainable materials are installed correctly and efficiently.

Regular inspection and maintenance during the construction process can help identify and address any issues with the materials promptly. This proactive approach can prevent potential problems from escalating and ensure that sustainable materials perform as intended.

CHAPTER 8: CONCLUSION AND FUTURE PERSPECTIVES

8.1 Recap of Sustainable Construction Materials and Techniques

Throughout this exploration, we have delved into a wide array of eco-friendly alternatives that contribute to a greener and more sustainable built environment. These materials and techniques have been carefully chosen for their minimal environmental impact, renewability, energy efficiency, and ability to support a circular economy.

From bamboo, hempcrete, and cork to recycled plastic, reclaimed wood, and sheep's wool, each sustainable material offers unique advantages and applications in construction. Bamboo stands out as a versatile and renewable building material with an impressive strength-to-weight ratio and fast growth cycles. Hempcrete, a lightweight and eco-friendly solution, provides excellent insulation and carbon sequestration properties. Cork serves as a natural insulation and fire-resistant material, harvested sustainably from cork oak trees. Recycled plastic and steel offer strength and durability with minimal environmental impact.

Reclaimed wood adds a touch of authenticity and sustainability to various construction elements. Lastly, sheep's wool presents a natural and renewable option for insulation and soundproofing.

The techniques employed in sustainable construction go hand in hand with the materials chosen. Building design plays a pivotal role, and principles such as energy efficiency, water conservation, waste reduction, and indoor environmental quality shape eco-friendly structures. Utilizing digital tools, Building Information Modeling (BIM), and Internet of Things (IoT) applications further optimize construction processes and improve efficiency.

8.2 Emerging Trends and Innovations in Eco-Friendly Building

As sustainable construction continues to gain traction, emerging trends and innovations are shaping the future of eco-friendly buildings. These advancements not only improve environmental performance but also enhance occupant comfort and well-being.

One of the key emerging trends is the increased adoption of mass timber and cross-laminated timber (CLT) in construction. Mass timber offers a renewable and carbon-negative alternative to traditional concrete and steel, as trees sequester carbon during their growth. Prefabricated timber components enable faster and more efficient construction, reducing the overall environmental impact.

Another notable trend is the integration of renewable energy sources into building design. Solar panels, wind turbines, and geothermal systems are becoming more prevalent, allowing buildings to generate their electricity and reduce reliance on non-renewable energy sources. The advancement of energy storage technologies also ensures a consistent power supply even during intermittent renewable energy generation.

Biophilic design is gaining popularity as well, focusing on

creating connections between occupants and nature within built environments. Incorporating natural elements, such as green walls, living roofs, and indoor plants, enhances indoor air quality, reduces stress, and fosters a sense of well-being.

Additionally, circular economy principles are increasingly influencing construction practices. Designing for deconstruction and material reuse, alongside the development of innovative recycling techniques, ensures that materials retain their value even after the building's life cycle ends.

Smart buildings equipped with advanced sensors and automation systems are another prominent trend in sustainable construction. These smart technologies optimize energy consumption, monitor indoor air quality, and adapt building systems based on occupancy patterns, ultimately leading to improved energy efficiency and resource management.

8.3 The Future of Sustainable Construction

The future of sustainable construction holds great promise, driven by continuous innovation and increasing awareness of environmental challenges. As climate change and resource depletion become more pressing concerns, sustainable construction will play an even more crucial role in mitigating the built environment's impact on the planet.

Materials science will continue to evolve, leading to the discovery of new sustainable materials with enhanced properties and applications. Researchers are exploring bioplastics, mycelium-based composites, and other bio-based materials that could further revolutionize the construction industry.

The integration of digital technologies, artificial intelligence, and robotics will streamline construction processes, reducing waste and improving efficiency. Drones, for instance, are already being used for site inspections and monitoring construction progress, enhancing safety and data collection.

Building designs will prioritize resilience and adaptation to climate change. Sustainable infrastructure, such as rainwater harvesting systems, green infrastructure, and natural ventilation, will become standard features to cope with extreme weather events and resource scarcity.

In the realm of energy, buildings will increasingly become energy-positive or net-zero energy, producing more energy than they consume. Smart grids and energy-sharing platforms will facilitate energy exchange between buildings, enabling a more robust and sustainable energy ecosystem.

Additionally, the concept of circular economy will be fully embraced in construction practices. Circular construction models will focus on designing buildings for disassembly and material reuse, promoting a regenerative approach to resource utilization.

The emphasis on occupant health and well-being will continue to grow, with biophilic design principles integrated into more buildings. Human-centric design, encompassing improved air quality, natural light, and thermal comfort, will enhance the overall living and working experience for occupants.

Sustainable construction is not just a passing trend; it is the future of the construction industry. With a vast array of eco-friendly materials and techniques available, along with ongoing advancements and innovations, the construction sector has the opportunity to lead the way towards a greener, more resilient, and sustainable built environment. As we continue to face global challenges, sustainable construction remains an essential pillar in building a more sustainable and thriving world.

BOOKS BY THIS AUTHOR

Aerogel As A Sustainable Construction Material: Towards Net Zero Carbon Emissions

This book is a comprehensive guide for anyone interested in learning about a groundbreaking material (Aerogel) and its potential to revolutionize sustainable building practices. The book provides a detailed overview of aerogel, its unique properties, and its applications in the construction industry. It explains how aerogel can be used to promote energy efficiency and reduce carbon emissions, making it an essential tool for achieving net-zero carbon footprints in construction projects.

Written by an expert in the field, this book covers a range of topics, including the science behind aerogel, its production and manufacturing, and its use in different building applications. It also includes case studies and real-world examples that showcase the practical applications of aerogel in sustainable construction.

The book is recommended for construction professionals, researchers, students, and people that are simply interested in learning about the latest developments in sustainable building practices. With its clear and accessible writing style, engaging content, and practical insights, the book is sure to inspire and inform readers about the potential of aerogel as a game-changing material in the pursuit of sustainable construction practices.

Building Sustainability: Practical Construction

Practices

Unlock the secrets to sustainable construction with this book. It is an invaluable resource for construction professionals, project managers, contractors, architects, and students seeking to integrate sustainable practices into their projects.

Discover the importance and benefits of sustainable construction as you explore sustainable design principles, materials selection, energy efficiency, green certifications, waste management, and more. Real-life case studies provide practical insights, while discussions on financial considerations and resilience in construction empower readers to make informed decisions.